Selected Papers from the 19th International Conference on Micro- and Nano-Technology for Power Generation and Energy Conversion Applications (Power MEMS 2019)

Selected Papers from the 19th International Conference on Micro- and Nano-Technology for Power Generation and Energy Conversion Applications (Power MEMS 2019)

Editors

Paweł Knapkiewicz
Rafał Walczak

MDPI • Basel • Beijing • Wuhan • Barcelona • Belgrade • Manchester • Tokyo • Cluj • Tianjin

Editors
Paweł Knapkiewicz
Faculty of Microsystem
Electronics and Photonics,
Wroclaw University of Science
and Technology
Poland

Rafał Walczak
Faculty of Microsystem
Electronics and Photonics,
Wroclaw University of Science
and Technology
Poland

Editorial Office
MDPI
St. Alban-Anlage 66
4052 Basel, Switzerland

This is a reprint of articles from the Special Issue published online in the open access journal *Energies* (ISSN 1996-1073) (available at: https://www.mdpi.com/journal/energies/special_issues/PowerMEMS_2019).

For citation purposes, cite each article independently as indicated on the article page online and as indicated below:

LastName, A.A.; LastName, B.B.; LastName, C.C. Article Title. *Journal Name* **Year**, *Article Number*, Page Range.

ISBN 978-3-03943-579-1 (Hbk)
ISBN 978-3-03943-580-7 (PDF)

Contents

About the Editors

Paweł Knapkiewicz is an Associate Professor at the Wrocław University of Science and Technology. He earned his Ph.D. in 2008 and started his professional career at the Faculty of Microsystems Electronics and Photonics as a research and teaching assistant. His research interests include microfluidics, especially microfluidic chip technology, microscale measurements of physical quantities, and the construction of multi-channel measurement systems for chemical microreactors, as well as the technology of cesium atomic microclock and micromechanical dosimeter of large doses of high-energy radiation. His ongoing research is focused on developing high-vacuum cesium/rubidium cell technology to obtain the world's first MEMS instrument, in which it will be possible to obtain the state of cold atoms. As part of his research, he broadly developed competences in the field of optical measurements, including optical interferometry, absorption spectroscopy of cesium and rubidium atoms (including "non-Doppler" saturation spectroscopy), infrared absorption spectroscopy, and laser cooling of atoms. The latest works are related to the development of precise optical systems and MEMS analytical instruments for nano satellites.

Rafał Walczak is the Dean of the Faculty of Microsystem Electronics and Photonics of Wrocław University of Science and Technology. He was the Head of the Division of Microengineering ad Photovoltaics and the Division of Microsystems. He is the leading researcher of many Polish and international (EU Framework Programmes) scientific projects related to the development of various devices utilizing micro/nanotechnologies and emerging technologies, including 3D printed energy harvesters. He is the author and co-author of over 200 scientific papers. During his career, he has attended more than 70 conferences and schools, he has had more than 40 oral communications at conferences. He also received many conference awards for outstanding oral and poster presentations and national prestigious awards including the Defender 2014 Award of Internationals Defence Industry Exhibition and 2014 Golden Laure of National Technical Organisation. He is a member of the Executive Board of the Polish Society of Sensor Techniques, a member of the International Steering Committee of Eurosensors Conference, and a member of MANCEF Foundation General Advisory Board.

Preface to "Selected Papers from the 19th International Conference on Micro- and Nano-Technology for Power Generation and Energy Conversion Applications (Power MEMS 2019)"

Thank you for completing the Special Issue "Selected Papers Form the 19th International Conference on Micro and Nanotechnology for Power Generation and Energy Conversion Applications (Power MEMS 2019). We would like to thank you for the high substantive level of the works presented at the Conference, which resulted in excellent publications included in this Special Issue.

The technical program of the Power MEMS 2019 consisted of 135 contributions, distributed in 67 oral and 68 poster presentations. Among them, three plenary sessions were presented by recognized experts, focused on novel fascinating applications of power MEMS for space exploration and various aspects of power generation and conversion. Focused sessions dedicated to micro-devices for space applications, energy harvesters for vehicle applications, and Polish–Singaporean cooperation were also proposed.

We want to thank you all for attending the Conference. It is the quality of your presentations and your passion to communicate with other participants that really made this conference a success.

The Power MEMS 2019 Conference took place in Kraków POLAND just before the global COVID-19 pandemic. Wishing you health and high-quality research, we hope that we will meet again soon and share our latest scientific discoveries.

Paweł Knapkiewicz, Rafał Walczak
Editors

energies

MDPI

Article

Thermoelectric Harvesting Using Warm-Blooded Animals in Wildlife Tracking Applications †

Eiko Bäumker *, Pascal Beck and Peter Woias

Laboratory for the Design of Microsystems-IMTEK, University of Freiburg, Georges-Koehler-Allee 102, 79110 Freiburg, Germany; pv.beck@t-online.de (P.B.); woias@imtek.de (P.W.)

* Correspondence: eiko.baeumker@imtek.de; Tel.: +49-76120367677

† This paper is an extended version of the authors' paper "Thermal energy harvesting through the fur of endothermic animals", published in 19th International Conference on Micro and Nanotechnology for Power Generation and Energy Conversion Applications (PowerMEMS), 2–6 December 2019 in Kraków, Poland.

Received: 24 April 2020; Accepted: 17 May 2020; Published: 1 June 2020

Abstract: This paper focuses on the design of an optimized thermal interface for a thermoelectric energy harvesting system mounted at endothermic animals. In this application scenario the mammal's fur reduces the heat flux from the animal's body through a thermoelectric generator (TEG) to the ambient air. This requires an adapted design of the thermal interface between TEG and body surface, to increase its thermal conductivity without harming the animal. For this purpose the thermal conductivity through a mammal's fur is determined with a specially designed heatsink. An analytical model is built to predict the resulting thermal resistances and is validated with experimental results for two different fur lengths. We show that an optimized design of the thermal interface reduces its thermal resistance up to 38% compared to a trivial design while lowering its weight for about 23%. It is found that the most important design parameter of such a thermal connector is the ability to slide into the fur.

Keywords: tracker; wildlifetracking; heat flux; thermal resistance; thermal conductance; thermal energy harvesting; endothermic animal; fur

---

1. Introduction

With the ongoing reduction of power consumption for sensor systems, energy harvesting (EH) techniques have become more popular in application fields like wildlife monitoring. Mobile trackers, attached to the desired animal, are used to gather important data about the host's habits and health status. Commercially available trackers such as the LifeTag^{TM} [1] already use energy harvesting techniques such as solar cells to prolong battery lifetime. Thermoelectric systems, harvesting energy from the temperature difference between host and environment, can be a more robust alternative with a more continuous energy income. However, in previous field experiments at sheep, a temperature difference of only 2.5 to 3.5 K was achieved between the surface of the animal's fur and the ambient air. With a reasonably sized thermoelectric generator (TEG)-air energy harvesting system the possible energy outcome in the range of 54 µW [2] is correspondingly low. This is a major issue for a typical wildlife tracking system that has to gather the position of the animal regularly. The global positioning system (GPS) commonly used for that task draws about 320 mW of power in active mode for a longer period of time (sometimes more than 30 s) [3]. This is not only a problem for energy harvesting systems, as it usually exceeds the energy incomes of those, but even in battery driven applications the power consumption of GPS heavily restricts the amount of positions that can be gathered when a long lifetime without changing batteries is desired.

An improvement of the energy income is challenging, as this normally means, for a thermoelectric EH system, an increased heat flux from the animal body into the TEG, and from there to the ambient air. Increasing the heat flux by using a TEG with larger area is only possible to some extent, as the growing weight and size restricts the attachment of arbitrary large devices. Thermal energy harvesting devices that are attached to human skin exist [4,5]. Design parameters to improve thermal conductivity and weight are present in this case, but this knowledge is not applicable for devices harvesting through the thick or thin fur of a mammal. A great difference is the significant lower thermal conductivity of fur compared to human skin. In addition to that, an optimized design of a thermal heat connector (THC) for animal fur will deviate from one made for bare skin. Although rough values for the fur's thermal conductivity can be obtained from existing literature [6–9], it is unclear how THCs should be designed to lower the resulting thermal resistance. In the first part of this paper, we explain the analytical model developed for an optimal design of a THC. Based on the findings from the model, several optimized THCs are fabricated. The following chapters concentrate on the experiment, which evaluates these designs in terms of thermal resistance, weight and penetration capability into the fur. Finally, we will compare the results of the experiment with the values expected from the model.

2. General Consideration for a Thermoelectric Energy Harvesting System Operated on an Animal

We focus on a thermoelectric energy harvesting system attached to an endothermic animal using the temperature difference between its host and the environment. Such a system can be described by an equivalent model with five simplified components, as shown in Figure 1. The animal is modeled as an inexhaustible heat source with an almost constant temperature T_{hot}. The fur and skin, together with a thermal heat connector to the TEG provide a first thermal resistance in the heat path between the animal's core temperature and the environment. Two additional thermal resistances are formed by the TEG itself and a heatsink mounted on top, which connects the system to the ambient air temperature T_{cold}.

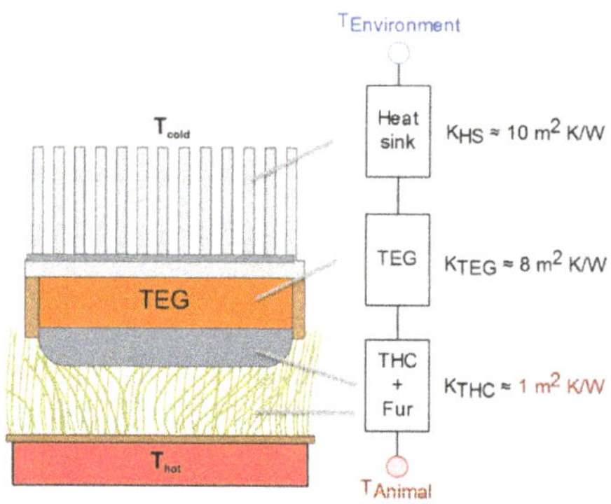

Figure 1. Equivalent circuit model for a thermoelectric harvesting system attached to an animal. In typical scenarios, the thermal conductivity of the fur is significantly poorer than for the other samples.

In our case, it is desired to maximize the output power of the TEG in such a system. Literature exists that describes the impacts of the thermal impedances, overall temperature difference and properties of the TEG on the output power thoroughly [10]. The choice of an appropriate TEG is critical for a good performance, but usually is a compromise between its thermal and electrical resistance as well as output voltage. A major challenge when only low temperature gradients are available for harvesting is the corresponding low output voltage of the TEG. With only a total temperature difference of about 10 K between the environment and the animal, the TEG's output voltage will be in the range of a few millivolts when optimizing for power output. This commonly leads to a poor usable electric power, considering that a necessary up-conversion of the TEG's output voltage to drive an attached electronic system is associated with high losses. It is therefore possible that a system focusing on a high TEG output voltage can deliver an overall higher usable power than one trying to solely optimize the power outcome of the TEG. Regarding the thermal impedance of the TEG, the user is frequently bound to TEGs of relatively high thermal conductivity which forces him to thermally adopt the heatsink and thermal heat connector.

An exemplary situation is described in the literature for energy harvesting from human body heat [5]. However, the main difference to highlight when harvesting from an animal is the lower thermal conductivity at the skin side. While the thermal resistivity of the human skin is around 0.05 m^2K/W [11], the corresponding value for an animal with a fur length of only 5 mm is between 1.25 and 0.33 m^2K/W [7]; i.e., much higher. As a result, the THC is the part with the largest potential of improvement in the chain of thermal resistances (see Figure 1) and has a great impact on the overall performance. One strategy could be to increase the size of a THC, thereby reducing its thermal resistance through an increased contact area. This approach is, however, not viable for two reasons: First, animal ethics dictate a maximum weight of, e.g., a collar below 6% of the animal's body mass [12]. In general, a minimal footprint and weight for systems attached to an animal are desired to lower the impact on the animal. Second, a THC with a large thermal mass will act as a thermal low pass filter in time. This means that rapid thermal transients are averaged and the temperature difference across the TEG is averaged in the same way. As a result, the thermal power and electrical power available, which is proportional to the square of the temperature difference, is reduced. It is therefore the aim of this study to find a design of a THC with a reasonable compromise of all these boundary conditions [13].

3. THC Design

The design of the THC is based on a typical heat sink made from aluminum; see Figure 2. The fins of this heatsink are cylindrical, as they should penetrate the fur of the animal. The idea is to reduce the overall thermal resistance of the fur by having a short-circuit thermal path through the fur, along a material with a much lower thermal resistance, here the fins.

When attaching the THC to the animal, the corresponding thermal path will split into three different sections: One part of the fur will be pushed between the fins and slightly compressed. The other part of the fur will be compressed between the animal's skin and the fins.

This scenario can be modeled with an equivalent circuit, as clarified in Figure 3. We assume that sliding the THC into the fur does not have an impact on the fur's thermal conductivity. Instead we consider that the fur evenly slides aside and fills the space between the cylindrical fins. With respect to that, the thermal path through the fins and through the fur to the rest of the harvesting system is of the same length l_{Fins}. Typically the THC will not be able to reach to the skin of the animal and a certain layer (thickness d_{skin}) of fur stays in between. Considering the animal as an inexhaustible heat source and neglecting other thermal resistances on the cold side, the constellation of mixed fur and fins can be expressed as two thermal resistances $R_{fins/fur}$ placed in parallel to each other. The remaining fur beneath adds a high resistance in series to that parallel circuit.

With the help of this equivalent circuit, the total combined resistance R_{THC} of fur and THC can be described by Equation (2) where l_{Fins} is the fin length and d_{Skin} the thickness of the fur that remains below the THC. The size of the THC and therefore the total area of harvesting is split into A_{Fur} and A_{THC}. A_{THC} corresponds to the total front face of the fins, whereas A_{Fur} relates to the area in between where only fur is present. The two thermal resistivity values are k_{Fur} for the fur and k_{THC} for the material of the fins.

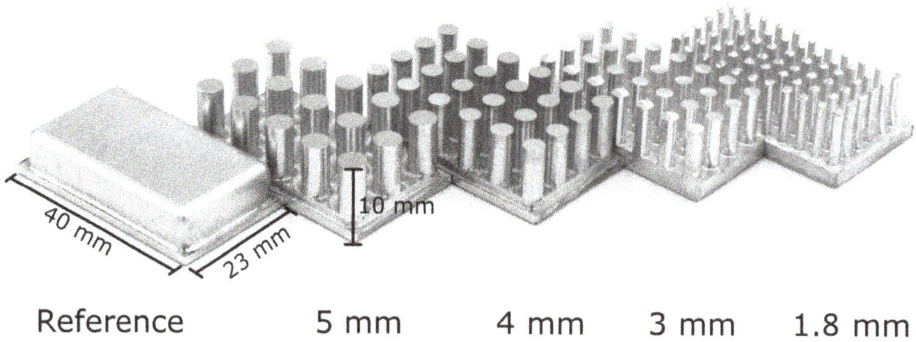

Figure 2. Photograph of the five different versions of the thermal heat connector (THC), here for sheep's wool, with variations in fin diameter and spacing. The fins height l_{Fins} is 10 mm for the experiment with the sheep wool and 3 mm for the goat fur.

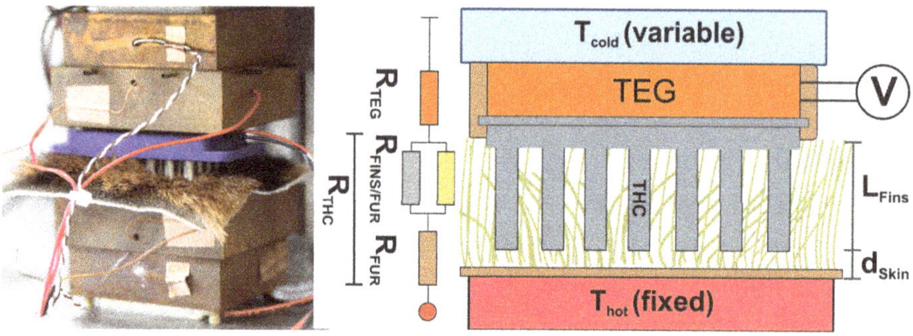

Figure 3. Left: Experimental set-up with the test device (THC and fur sample) put between two temperature-controlled plates. The THC with a fin length of 10 mm was used with the short fur for demonstration purposes. Right: Schematic and equivalent circuit of the THC at the fur. The fins of the THC act as a parallel thermal resistance R_{Fins} to the thermal resistance R_{Fur} of the displaced fur. The remaining fur between THC and skin is represented as a thermal resistance R_{Fur} in series.

For a single thermal resistance R_{Therm}, Equation (1) is used to describe its value as a function of the thermal conductivity of its material (k), its surface area (A) and the length of the heat path through the resistance (L):

$$R_{Therm} = \frac{L}{A} k^{-1} \tag{1}$$

Using standard equations for the parallel and serial electrical circuit of resistors, the following Equation (2) holds for the combination of the fur and THC:

$$R_{THC} = \frac{l_{Fins}}{A_{Fur} k_{Fur} + A_{THC} k_{THC}} + \frac{d_{Skin}}{(A_{Fur} + A_{THC}) k_{Fur}} \tag{2}$$

It is worth mentioning that the parameter k_{THC} for aluminum of 237 W/(m K) is four to five orders of magnitude greater than k_{Fur} with 0.04–0.15 W/(mK). For that reason, the first term of Equation (2) is small even with a tiny face area A_{THC}. As a result, the overall resistance R_{THC} is determined by the second fraction. Decreasing the second part of the equation can be done by two options: either by minimizing d_{Skin}, the distance between the fins and the bare skin, or by increasing the fins' amount (A_{THC}) of the total harvesting area $A_{Fur} + A_{THC}$. A third option, increasing the total harvesting area $A_{Fur} + A_{THC}$ itself, is normally not desired. In the real world, d_{Skin} and A_{THC} can not be optimized without affecting each other in a negative manner. For example, a dominating A_{THC} would on one hand decrease the thermal resistance, but would compress a lot of fur beneath the fins on the other, thereby increasing the overall distance to the skin. Care must be taken that there is enough room for the fur to slide in between the fins to not negatively affect d_{Skin}. Accordingly, the fin diameter and spacing between the fins need to be reasonable.

The resulting distance d_{Skin} to the skin will be also affected by the length l_{Fins} of the fins. A high length allows the fins to get closer to the skin, as the bottom part of the THC will not compress the fur beneath. However the length can not be arbitrarily high. The fins have to be long enough to penetrate deep enough into the fur, but any additional length would leave space between the fur's surface and the THC's bottom. This gap is detrimental in its effect, as it will provide a path for heat dissipation to the surrounding air, bypassing the heat flux through the TEG.

Therefore, an appropriate THC design should focus on the penetration capability when typical materials with high thermal conductivity are used for the THC itself. The penetration capability is expected to be dependent on the shape of the fins. Rounded or pointed cylinders have a greater chance to push the fur aside, whereas fins with a flat top more likely compress the fur beneath. In addition to that, the diameter and spacing of the fins has to be taken into account: While larger diameters would be desirable in order not to harm the animal's skin, their larger front area tends to compress the animals' fur to a higher degree. This will increase the thermal resistance as a detrimental effect. Besides that, it is always possible to lower the overall resistance via a greater harvesting area with the disadvantage of a larger weight and size of the THC.

For our experimental setup, we are using only designs with fully cylindrical fins which may not be the optimal shape with the best penetration capability, but are taken here due to a simpler manufacturing process. As the new design is compared to a trivial design, here an aluminum plate with a planar surface (see Figure 2), it is still possible to derive whether the penetration capability has an impact on the overall resistance.

4. Experimental Setup

An experimental setup was built to verify the validity of the developed model and to get an approximate value of the thermal resistance for each of five different THC designs. To cover a wide range of fur types, we adopted the THC designs for two extreme cases: The first set was designed for an animal with long and thick fur, whereas the second set is representative of animals with very short fur. It can be expected that the behavior of other animal fur lies in between the gathered results. Due to regulations and market availability, the wool of a merino sheep was taken for the long-fur set, while the fur of a domestic goat was used for the short-fur set.

Each of the two sets consist of four different THCs which differ in the diameter of the fins and in the space in between. They are named according to their fin diameter, as shown in Figure 2. The length l_{Fins} of the fins in each set corresponds to the fur length it was made for. For the set intended for the sheep fur, an equal fin length l_{Fins} of 10 mm was chosen. The fin length for the set with the goat fur was 3 mm. A photo of the fabricated THCs is shown in Figure 2.

The basic setup for the experiment is shown in Figure 3: The THC is put into a housing together with the TEG that is used to determine the thermal flux. This device is then placed between a hot plate with a fixed temperature, covered by a piece of fur, and a temperature-controlled cold plate in close contact with the TEG. The test device is pressed gently into the fur by the upper plate, as demonstrated in Figure 3. The temperature of the bottom plate is fixed to 33 °C, based on temperature measurements at the skin of a sheep in [9]. The upper plate's temperature is varied between 20 °C and 30 °C, resulting in a total temperature gradient ΔT_{Env} of 3–12 K. The measurements were taken in a temperature controlled room at 21 °C and a relative humidity of 60%.

After each adjustment of the temperature, 20 min of waiting was considered before the output voltage of the TEG was measured. The temperature gradient ΔT_{TEG} at the TEG can be derived from

that the thermal path cannot be improved as much as this is possible in the case of a long fur and fin length.

Two results of the experiment are worth being pointed out. First, the THC with a fin diameter of 1.8 mm performs bad at both fur lengths, despite having the smallest fin diameter. A possible explanation is the THC's minimal space between the fins among all the tested ones. Second, the THC with a fin diameter of 3 mm in the sheep's fur has a remarkably higher thermal resistance compared to the other. This could be an indication that the results may have been influenced by parameters that have not been recorded, like slight variations in the force that is applied to the TEG-THC combination.

Besides the fin diameter and spacing between the fins, an important parameter is the capability of a THC to slide into the fur, summarized with the other parameters in Table 2. It can be seen that a good sliding into the sheep's wool is observed for the 4 mm diameter THC. Here, the major part of the fur slides between the fins and is not compressed beneath. This could be due to the fact that this THC has the second highest spacing between the fins and exhibits a small total front face area. Despite this small contact area, the 4 mm THC works best at both fur lengths. It is worth mentioning that the lateral fin spacing and front surface of the 4 mm and 5 mm THCs are similar, but the resulting thermal resistances differ significantly. This could be an indication for the importance of the penetration capability, as the 5 mm diameter fins are not able to appropriately displace the fur; they instead compress it beneath. Other fin designs may behave differently; e.g., a rounded shape on top of the fin.

To check the results for plausibility, we derive the theoretical distance d_{Skin} between fur and fins from Equation (2). The thermal resistance R_{THC} is directly taken from the measured results, whereas we define k_{Fur}, as 0.1 ± 0.04 W/(mK) taken from [6] as an estimation for the fur's thermal resistivity. The thermal resistance of the THC, k_{THC} is fixed to 237 W/(m K). It is reasonable to expect that the calculated d_{Skin} for each THC should be in the range of only a few mm. Table 3 indeed states that the corresponding d_{Skin} is reasonable, but its improvement from 5 mm of the reference down to a d_{Skin} of 3.7 mm for an optimized THC at the sheep is too small compared to the fin length of 10 mm. An improved setup could reveal here more about the influence of the fin length and applied force on the THC and the corresponding compression of the fur beneath.

Table 3. Computed distance between fins and skin of the animal with Equation (2). The distance was calculated from the measured thermal resistance and a the theoretic value of k_{Fur} and k_{THC}.

Fin Ø	Sheep d_{Skin}/mm	Goat d_{Skin}/mm
Reference	5.1 ± 2.0	2.2 ± 0.9
1.8 mm	5.0 ± 2.0	3.8 ± 1.5
3 mm	7.0 ± 2.8	3.5± 1.1
4 mm	3.7 ± 1.5	2.3± 0.9
5 mm	4.2 ± 1.7	2.8± 1.1

In general, the difference of the thermal resistance between reference and optimized THC is greater for the long fur than it is for the short which we trace back to the longer thermal path that can be optimized at longer fur lengths. The measurement of the fur's thermal resistance is challenging, as it is dependent on various parameters like humidity, fur length and type. In addition to that the THC itself has various design parameters, of which the penetration capability especially cannot be accurately determined. A relative comparison, as done in this paper, between several THC designs circumvents a lot of the mentioned errors. The gathered thermal resistances of the THC are in the expected range, which is calculated as 81 to 22 K/W for a length $l_{Fins} = 3$ mm of the goat fur. For the sheep fur with a length at $l_{Fins} = 10$ mm the measured thermal resistance is below our expectation of 72 to 272 K/W. In the experiments, a value between 40 and 72 K/W was found. This could be an indication for an overestimated fur length and thereby l_{Fins} in the setup.

Overall, the experiment demonstrates that a THC design with fins can lower the overall thermal resistance by up to 38% as well as reduce the weight in comparison to a planar bulk design. According to Equation (2), the diameter of the fins should not play a significant role. Instead, the remaining fur between the tip of the fins and the skin will dominate the total thermal resistance of the interface. The experiments are in accordance to that assumption as they do not show a clear trend of thermal resistance dependency on the fin diameter. Additionally, the results do not give a clear indication that the spacing between the fins has a great impact. Both of them, however, have an impact on the ability of the THC to penetrate the fur which correlates to the thermal resistance. We assume that a good penetration capability would bring the THC closer to the skin, reducing d_{Skin} in Equation (2).

6. Summary and Conclusions

We investigated improving the heat flux through a thermoelectric energy harvesting system used at endothermic animals. Consisting of a thermoelectric generator (TEG), heatsink and thermal heat connectors (THC), we found that the THC offers the largest potential for an improved heat flux and therefore power outcome of such a system. A new and adopted design for a THC was developed, based on that of a typical heatsink with cylindrical fins, aiming to improve the heat flux while maintaining minimal weight and dimensions. Concerning the thermal resistance of the THC, our derived mathematical model states that a good design should have fins that get as close as possible to the animal's skin. The fin diameter itself plays only a minor role according to the model equations developed in this research.

We have then verified the results predicted by the model in an experiment. Different designs of THCs were fabricated and their thermal resistances were determined with the fur of a long and short haired animals, respectively. The experiments reveal that one major influence factor on the thermal resistance is only indirectly covered by the formulated model equation: the penetration capability of the THC design is highly dependent on shape and width of the fins, and on the spacing in between. While fins with smaller diameters slide better into the fur, a large enough spacing between the fins becomes more important to give the fur room for mingling with the THC and to get as close as possible to the skin. The 4 mm pile diameter design of the THCs has achieved a thermal resistance that is 38% better than the reference design and has less weight compared to the same.

As far as we know, this is the first time that the thermal resistance of fur with an attached heat connector has been examined. The results can be seen as a starting point toward obtaining THCs with lower weight and improved heat flux. Besides the results of this paper showing that the penetration capability is one important parameter, other environmental effects like humidity or differences in the structure of the fur may heavily influence the overall thermal resistance of the connection between the animal and device. Field experiments to reveal such effects are on the way and may help in simulating such a device.

The outcome can be useful for energy-autonomous animal tracking systems that are put on an endothermic animal to harvest energy directly from their hosts. With the chance of an improved energy income or smaller size, this could be used to prolong battery lifetime or solely run the system from harvested energy. A reduced weight will also allow the user to attach these systems to smaller animals where the maximum allowed 6% of the body weight is restricting thermal harvesting. In fact, a system built out of two of the proposed THCs and two corresponding heatsinks will have a weight around 120 g, excluding electronics and the collar itself. This is in the range of commercially available GPS-trackers like the Vertex Plus Collar, with a total weight of at least 470 g [14]. The benefit we expect from this system is, however, its longer operational lifetime, as no exhaustible energy storage is used.

Author Contributions: Conceptualization, Validation, Formal Analysis, Writing, P.W. and E.B.; Preparation, Experimental data acquisition, P.B. and E.B. All authors have read and agreed to the published version of the manuscript.

Funding: The article processing charge was funded by the Baden-Wuerttemberg Ministry of Science, Research and Art and the University of Freiburg in the funding program Open Access Publishing.

Conflicts of Interest: The authors declare no conflict of interest.

References

1. *LifeTag^{TM}*; Cellular Tracking Technologies: Rio Grande, NJ, USA, 2019.
2. Woias, P.; Schule, F.; Bäumker, E.; Mehne, P.; Kroener, M.;Thermal Energy Harvesting from Wildlife. *J. Phys. Conf. Ser.* **2014**, *557*, 012084. [CrossRef]
3. Kjærgaard, M.B.; Langdal, J.; Godsk, T.; Toftkjær, T. Entracked: Energy-Efficient Robust Position Tracking for Mobile Devices. In Proceedings of the 7th International Conference on Mobile Systems, Applications, and Services (MobiSys 2009), Kraków, Poland, 22–25 June 2009.
4. Settaluri, K.T.; Lo, H.; Ram, R.J. Thin Thermoelectric Generator System for Body Energy Harvesting. *J. Electron. Mater.* **2012**, *41*, 984–988. doi:10.1007/s11664-011-1834-3. [CrossRef]
5. Thielen, M.; Sigrist, L.; Magno, M.; Hierold, C.; Benini, L. Human body heat for powering wearable devices: From thermal energy to application. *Energy Conv. Manag.* **2017**, *131*, 44–54. [CrossRef]
6. Cena, K.; Clark, J. Heat balance and thermal resistances of sheep's fleece. *Phys. Med. Biol.* **1974**, *19*, 51. [CrossRef]
7. Cena, K.; Monteith, J.L. Transfer Processes in Animal Coats. II. Conduction and Convection. *Proc. R. Soc. Lon. Ser. B Biol. Sci.* **1975**, *188*, 395–411.
8. Cena, K.; Monteith, J.L. Transfer processes in animal coats. I. Radiative transfer. *Proc. R. Soc. Lon. Ser. B Biol. Sci.* **1975**, *188*, 377–393.
9. Gatenby, R.M.; Monteith, J.; Clark, J. Temperature and humidity gradients in a sheep's fleece. I. Gradients in the steady state. *Agric. Meteorol.* **1983**, *29*, 1–10. doi:10.1016/0002-1571(83)90071-7. [CrossRef]
10. Spies, P.; Pollak, M.; Mateu, L. *Handbook of Energy Harvesting Power Supplies and Applications*; CRC Press: Boca Raton, FL, USA, 2015.
11. Leonov, V.; Torfs, T.; Fiorini, P.; Van Hoof, C. Thermoelectric converters of human warmth for self-powered wireless sensor nodes. *IEEE Sens. J.* **2007**, *7*, 650–657. [CrossRef]
12. Sodeikat, G.; Neubeck, K.; Schmidt, J. Verhaltensuntersuchungen an juvenilen Birkhühnern nach Anbringung von unterschiedlich schweren Halsbandsendern—Eine Pilotstudie. *NNA Sonderheft* **2008**, *19*, 43–46.
13. Moser, A.; Erd, M.; Kostic, M.; Cobry, K.; Kroener, M.; Woias, P. Thermoelectric Energy Harvesting from Transient Ambient Temperature Gradients. *J. Electron. Mater.* **2012**, *41*, 1653–1661. [CrossRef]
14. *Vertex Plus-Datasheet*; VECTRONIC Aerospace: Berlin, Germany, 2018.

energies

MDPI

Article

Separation-Independent Wearable 6.78 MHz Near-Field Radiative Wireless Power Transfer using Electrically Small Embroidered Textile Coils

Mahmoud Wagih *, Abiodun Komolafe and Bahareh Zaghari

School of Electronics and Computer Science, University of Southampton, Southampton SO17 1BJ, UK; a.o.komolafe@soton.ac.uk (A.K.); bahareh.zaghari@soton.ac.uk (B.Z.)
* Correspondence: mahm1g15@ecs.soton.ac.uk; Tel.: +44-2380-593-234

Received: 27 November 2019; Accepted: 17 January 2020; Published: 21 January 2020

Abstract: Achieving a wireless power transfer (WPT) link insensitive to separation is a key challenge to achieving power autonomy through wireless-powering and wireless energy harvesting over a longer range. While coupled WPT has been widely used for near-field high-efficiency WPT applications, the efficiency of the WPT link is highly sensitive to separation and alignment, making it unsuitable for mobile systems with unknown or loose coupling such as wearables. On the other hand, while ultra-high frequency (UHF) and microwave uncoupled radiative WPT (0.3–3 GHz) enables meters-long separation between the transmitter and the receivers, the end-to-end efficiency of the WPT link is adversely limited by the propagation losses. This work proposes radiative WPT, in the 6.78 MHz license-free band, as a hybrid solution to separation-independent WPT, thus mitigating the losses associated with coil separation. Resonant electrically small antennas were fabricated using embroidered textile coils and tuned using L-matching networks, for wearable WPT. The antenna's efficiency and near-fields have been evaluated numerically and experimentally. The proposed WPT link achieves a stable forward transmission of $S_{21} > -17$ dB and $S_{21} > -28$ dB, independent of coil separation on the XZ and XY planes respectively, in a 27 m^3 volume space. The presented approach demonstrates the highest WPT link efficiency at more than 1-m separation and promises higher end-to-end efficiency compared to UHF WPT.

Keywords: coils; wireless power transfer; electrically small antennas; e-textiles; internet of things; wireless energy harvesting

---

1. Introduction

Meeting the energy demands of a connected world is an increasingly growing challenge. Energy harvesting and scavenging, down to µW levels, is increasingly seen as a solution to the foreseen shortage in energy storage devices. To illustrate, while low power energy harvesting, transport and storage techniques are limited to applications such as sensor nodes and assisted living in smart cities, achieving power autonomy in ubiquitous internet of things devices reduces their demand for energy storage devices. Therefore, as fewer devices will require batteries, the competition for Lithium, a limited resource, can be minimized. Potentially allowing batteries to be exclusively used in high-end applications such as electric vehicles and smart grids [? ?].

Body-centric energy harvesting has been extensively explored. Piezoelectric [?], thermoelectric [?] and triboelectric energy harvesters [?], have been reported for harvesting energy generated from the human body. Furthermore, ambient harvesters such as solar cells [?], as well as microwave and mmWave rectennas have been designed on textile substrates [? ?], towards achieving power autonomy of wearable devices. A key limitation of such harvesters is their reliance on certain body-positions where the energy is generated such as the foot [?] or friction points [?]. Thus, the challenge of on-body position-independent energy transfer on textiles remains unsolved. A hybrid approach between body-centric energy harvesting and full-reliance on energy storage is wireless charging of e-textile electronics using inductive power transfer or magnetic resonance using flexible coils [? ?]. In [?], a switched-reluctance generator has been proposed as an on-bicycle power supply for textile-based sensing and communication through wireless power transmission from the bicycle to the body.

Wireless Power Transfer (WPT) is an increasingly popular method of delivering power to an electronic system [? ?]. While established applications of inductive WPT are omnipresent in the consumer electronics market and electric vehicle charging, the next generations of WPT devices will be based on resonant coupling to allow for increased separation between the transmitter and the receiver, in addition to non-coupled methods, through far-field propagation of WPT-specific waveforms. A particularly interesting platform for WPT is wearable e-textile applications, where establishing a real connection to the power supply is not feasible, and traditional energy harvesters require complex fabrication techniques to be integrated in textiles and flexible materials, or require specific on-body positioning [? ? ? ?]. Therefore, delivering power wirelessly to a wearable device represents a simple way of charging wearable nodes [?].

A hybrid approach combining sub-100 MHz WPT and radiative WPT could be adopted to mitigate the propagation losses and simplify the matching network design, improving the maximum achievable rectification efficiency. This work proposes radiative WPT using impedance-matched resonant electrically small coils, at 6.78 MHz. The proposed frequency band and antenna design reduce the complexity and cost of designing a matching network and maintain a stable WPT efficiency regardless of the coil's separation. This paper discusses the motivation behind the radiative WPT at 6.78 MHz, the fabrication of fully textile coils for WPT, theoretically and experimentally discusses the limitations of electrically small coils for WPT, and evaluates the performance of the system under different operation conditions.

2. Radiative 6.78 MHz WPT and System Architecture

2.1. WPT: The Existing Approaches

The broadband spectrum of WPT research can be summarized in multiple tracks, based on their frequency of operation and subsequently the application. Figure **??** shows examples of the most-reported WPT approaches. At lower frequencies, resonant and non-resonant (inductive WPT) coupled coils are used for high-efficiency WPT between two or more coils in consumer electronics and EV applications. Such technique is capable of handling kilowatts of power and is well-characterized and regulated [?]. Strongly coupled magnetic resonance, achieved using a tuned intermediate resonator between the transmitter and the receiver, has been demonstrated with high efficiencies at a few meters separation between the coils [?].

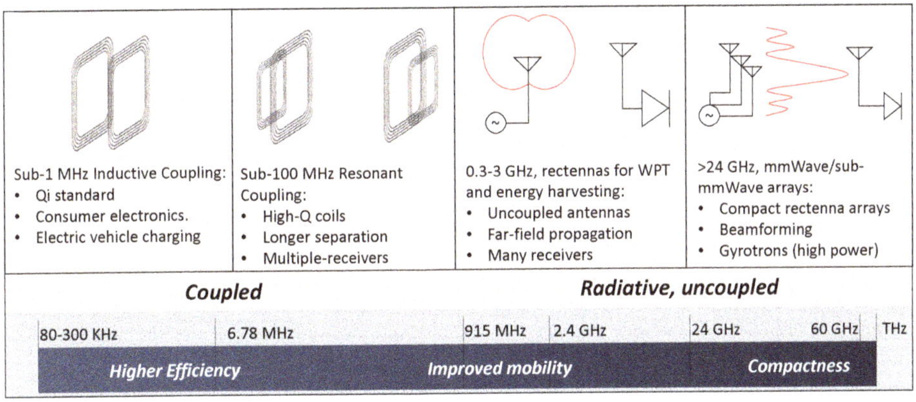

Figure 1. The current WPT research spectrum: sub-MHz Qi-standard WPT [?], coupled magnetic resonance [?], rectennas for UHF WPT [? ?], mmWave, sub-mmWave and THz optical rectennas [? ?].

In coupled WPT, efficiencies higher than 80% have been widely reported in wearable applications [?], nevertheless, most closely coupled resonant systems are not efficient beyond 20 cm of separation s [?]. This is attributed to the WPT efficiency being proportional to $1/s^6$ [?], as opposed to $1/s^2$ in free-space propagation. Therefore, coupled WPT approaches are limited in the distance achieved. In addition, at very close separations, magnetic resonance WPT systems are prone to over-coupling and require reconfigurable tuning circuitry for re-tuning the coils in over-coupling and under-coupling regions [? ?].

At UHF and microwave bands, WPT is implemented using electromagnetic radiative rectennas, commonly formed of a resonant distributed elements antenna (i.e., not requiring lumped components matching) and a rectifier. Based on their application: harvesting ambient radio-frequency signals or receiving power from a dedicated feed; multiple antenna and rectifier designs have been proposed [?]. The key advantages of this approach is the absence of a coupling link between the transmitter and the receiver, allowing more spatial freedom. Furthermore, this technique allows many receivers to be powered using the same source. At millimeter-wave bands and beyond (THz and infrared) large antenna arrays and nano-rectennas are used in power-beaming applications with power handling capabilities in the kW regime [?].

The far-field WPT end-to-end efficiency is significantly suppressed by the propagation losses, given by the Friis formula (1), where P_{TX} is the transmitted power, G_{TX} and G_{RX} are the transmitter and receiver gain, respectively, λ is the free-space wavelength and d is the separation between the transmitting and receiving antennas.

$$\textit{Friis Path Loss}\,(\text{dB}) = P_{TX} + G_{TX} + G_{RX} + 20\log_{10}(\frac{\lambda}{4\pi d}) \quad (1)$$

For example, in ambient RFEH, the reported power densities rarely exceed µW/cm due to the separation between the base-station and the rectenna [?]. In addition, indoor RF power densities are adversely reduced due to the absence of a direct line-of-sight, as well as the multi-path effects, resulting in over 10 dB reduction of the measured power densities [?].

It is observed from (1) that the power at the receiver is inversely proportional to the distance and frequency squared ($\lambda = 3 \times 10^8/\text{frequency}$). This implies less than 1% end-to-end forward transmission ($S_{21} < -20$ dB), for separations over 1 m at UHF bands. While antenna arrays and beamforming can be used to improve the power delivered to the receiver without increasing the transmitted power (through higher G_{TX}), the maximum transmitter gain and power are limited by regulatory bodies such as the Federal Communications Commission (FCC) under the Equivalent Isotropic Radiated Power (EIRP) limit.

An additional challenge at higher frequencies lies in the RF to DC converter, the rectifier. While rectifying antennas (rectennas) in theory achieve up to 100% efficiency, the voltage drop across the diode, and the reflection at the diode's input imply that the RF-DC efficiency is bottle-necked, especially at power levels below −10 dBm (100 μW). Furthermore, the design of a matching network becomes more complicated at higher frequencies due to the necessity of characterizing the diode's packaging parasitic at the frequency of interest.

To illustrate the WPT challenges at microwave frequencies, the propagation losses at UHF bands have been calculated using (1) at 1 m separation, demonstrating the logarithmic increase in propagation losses with frequency. Furthermore, a low-barrier Schottky diode (Skyworks SMS7630), widely used in rectenna-design studies, has been simulated using harmonic balance in Keysight Advance Design Systems (ADS) to calculate its maximum achievable RF-DC efficiency assuming ideal impedance matching, at 0 dBm, using a 50 Ω source. Figure ?? shows the decrease of the maximum RF-DC efficiency at microwave bands along with the logarithmic increase in propagation losses with frequency.

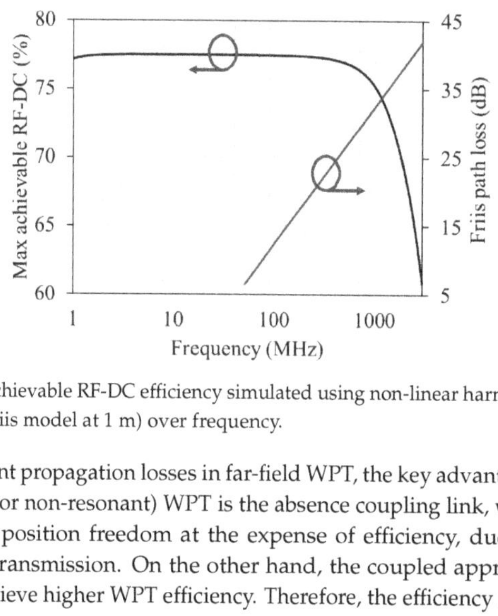

Figure 2. Maximum achievable RF-DC efficiency simulated using non-linear harmonic balance, and the propagation losses (Friis model at 1 m) over frequency.

Despite the significant propagation losses in far-field WPT, the key advantage to far-field rectennas over inductive (resonant or non-resonant) WPT is the absence coupling link, where a radiative system is able to achieve more position freedom at the expense of efficiency, due to relying on far-field propagation for power transmission. On the other hand, the coupled approaches rely on the coils' mutual inductance to achieve higher WPT efficiency. Therefore, the efficiency is generally proportional to $1/s^6$ [?].

2.2. *Radiative Near-Field WPT*

Motivated by overcoming the path losses and RF-DC inefficiencies observed in Figure ??, and overcoming the inverse-square relation with the propagation distance, selecting a sub-10 MHz WPT carrier frequency becomes an apparent choice. Where the radiative nature of the proposed approach allows coupling-free WPT, multiple receivers support, and the potential to achieve higher separation distances. The 6.78 MHz Industrial Scientific and Medical (ISM) band has been selected as the design frequency in this work. At 6.78 MHz the specified limits by the FCC are a maximum electric field of 30 μV/m (−45.7 dBm), measured at 30 meters from the source. Moreover, in certain biomedical applications, the FCC has approved a waiver of the limit for up to 200.2 μV/m (−29.2 dBm) for a wirelessly powered device demonstrating minimal interference with the surroundings [?]. While such low power limits constrain this WPT technology to ultra-low power applications, multiple systems have been presented with sub-μW power consumption for implants and biomedical applications, which justify the need for separation-independent WPT even if it is only achievable in the μW-range [?].

The proposed system, shown in Figure **??**, uses impedance-matched coils to radiate at 6.78 MHz; acting as electrically small antennas. A sub-100 MHz carrier benefits from lower propagation losses and free positioning, compared to UHF bands and coupled WPT, respectively. As the focus of this work is investigating the efficiency of the WPT link, the WPT system (the coils and their matching network) is evaluated as a black box using its scattering-parameters (s-parameters) matrix. The forward transmission (S_{21}) is the main parameter measured in this work, which can be used to calculate the WPT efficiency using $\eta_{WPT}=|S_{21}|^2$. Using s-parameters for evaluating WPT through an electromagnetic black box or for near-field applications has been previously widely used to simplify the problem where the electromagnetic medium is difficult to model (such as on-body operation) [?]. The transmitter and receiver coils and their matching network are symmetric as both are matched to 50 Ω transceivers.

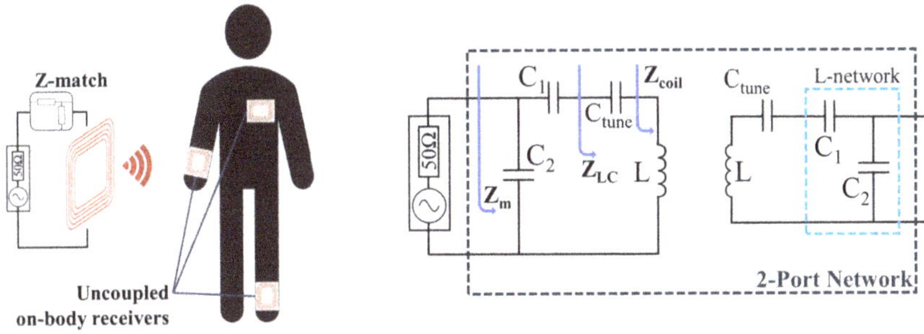

Figure 3. The proposed uncoupled radiative WPT: (**a**) system architecture, (**b**) 2-coil system schematic and impedance matching approach.

3. Textile WPT Coils Design and Fabrication

The main area-consuming component in a WPT system is the coil or the antenna, where the size can affect the maximum achievable WPT efficiency. Therefore, the coils are the only component from Figure **??** implemented on textile, as the capacitors could be realized using low footprint surface mount components or an on-chip capacitor bank integrated with the wireless power transmission or sensing circuitry on the same system-on-chip. The geometry of the coils has been based on a standard planar coil geometry for ease of fabrication and design.

Multiple fabrication techniques have been proposed to realize WPT coils on textiles [? ?], with the main affected coil parameter by the fabrication technique being the series resistance. For example, conductive inks normally have a sheet resistance of more than 1 Ω/square. Therefore, silk-coated copper Litz wires are chosen as the conductive material. Given their 40 μm thickness, they are expected to have minimal impact on the fabric's flexibility and hence the user's comfort. The coils are embroidered to a 350 μm-thick polyester-cotton textile substrate using an automated embroidery machine using a CAD model, the coil has been design using EAGLE, a commercial PCB design package. Figure **??** shows the layout and photograph of textile coils. A cross-section scanning-electron microscope photograph of the coils, showing the Litz conductors is shown in Figure **??**.

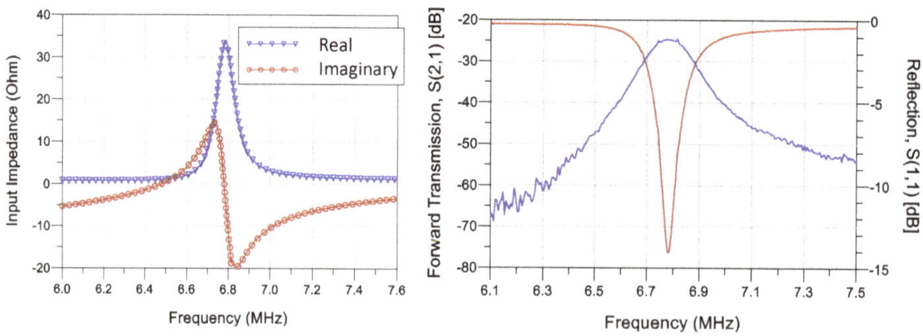

Figure 7. Measured coil input impedance (**Left**) and two-port s-parameters of symmetric matched coils at 10-cm separation (**right**).

4.2. Analytical Antenna Efficiency

The radiation efficiency of an antenna represents the ratio of the power radiated in space, to the power accepted from the source excluding impedance mismatch losses. An electrically small loop antenna can be considered using its equivalent circuit formed of a series inductor, the loop's Ohmic resistance R_{loss}, representing the power dissipated in the conductors in the form of heat, and a radiation resistance R_{rad} representing the power radiated to space. The radiation efficiency of an antenna is given by the ratio of R_{rad} to the total antenna's resistance [?].

$$\eta_{radiation} = \frac{R_{rad}}{R_{rad} + R_{loss}} = \frac{R_{rad}}{R_{total}} \tag{7}$$

To overcome the difficulty of measuring R_{rad} and R_{loss}, the total resistance R_{total} can be analytically calculated from the antenna's Quality factor Q and the loop's inductance L (**??**). Q can be calculated from the bandwidth using $Q = \frac{f_0}{BW}$. The bandwidth BW defines the frequency bandwidth where $S_{11} < -6.99$ [?]. To calculate the radiation efficiency, R_{rad} needs to be estimated using (**??**).

$$R_{total} = \frac{\omega L}{Q} \tag{8}$$

$$R_{rad} = 60\pi^2(ka) \int_0^{2ka} J_2(x)dx \tag{9}$$

ka represents the size of an electrically small loop antenna, where a is the loop's radius and $k = 2\pi/\lambda_0$. A standard approximate term for a single-turn loop antenna's radiation resistance R_{rad} is given by (**??**), as a function of the loop's area [?]. As the coil in this work resembles a square, s is given by $\frac{2l}{\pi}$. For a multi-turn loop, N is the number of coil turns.

$$R_{rad} = \frac{\mu_0 c}{6\pi}(\frac{2\pi}{\lambda})^2 \times (NS)^2 \approx \frac{31171(NS)^2}{\lambda^4} \tag{10}$$

From (**??**), the radiation resistance was calculated to be 0.37 mΩ. To calculate the radiation efficient, R_{total} has been calculated using (**??**) from the measured $S_{11} < 7.99$ dB bandwidth of 83.13 KHz and 6.78 MHz center frequency to be R_{total} = 5.278 Ω. Using (**??**) the antenna's radiation efficiency $\eta_{radiation}$ is calculated to be −41.5 dB.

4.3. Wheeler Cap Antenna Efficiency Measurement

The Wheeler Cap method, proposed to measure the radiation efficiency without the need for an anechoic test range, has been used to measure the $\eta_{radiation}$ of the matched textile antenna. The method can be summarized in covering the radiating antenna using a "cap" shorted to the antenna's ground, hence confining the radiated power to the antenna's ground return path. Therefore, the measured antenna's input impedance no longer represents R_{total} and is only composed of R_{loss}. The input impedance measurement is then repeated in absence of the cap to obtain R_{total}, R_{rad} is then calculated from the difference between R_{total} and R_{loss} [?].

To improve the repeatability of the measured efficiency the input impedance was measured twice in the presence and absence of the cap. A VNA was used to measure the input impedance in a similar manner to that described in the previous section. Figure **??** shows the measured antenna's response in the presence and absence of the Wheeler Cap, the difference in the reflection coefficient magnitude observed when the cap is present represents the power radiated by the antenna.

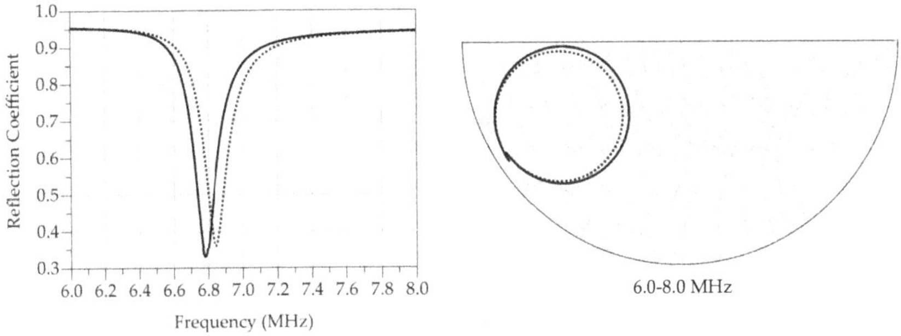

Figure 8. One-port antenna measurements in the presence (dashed) and absence (solid) of the Wheeler Cap: (**Left**) magnitude response, (**Right**) complex input impedance.

From the Smith chart in Figure **??**, the average measured $R_{total} = 26.584\ \Omega$, and $R_{rad} = R_{total} - Rloss = 1.514$, the average calculated radiation efficiency $\eta_{rad} = 4.33\% = -13.66$ dB.

5. Performance Evaluation

5.1. Coil Near-Field Simulation and Measurement

The near-field magnetic-field emissions from the coil were measured at various separations using a near-field probe. Should the proposed coils be used in the far-field region, multiple methods were reported in the literature to evaluate the far-field radiation properties based on near-field scans [?].

A VNA has been used to measure the H-field as the forward transmission between the coil and an H-field probe. The probe used is the Rhode and Schwarz RS H400-1, providing the average magnetic field across its 25 mm radius. While the probe is designed for signals higher than 30 MHz, the manufacturer's parameters show additional attenuation of approximately -10 dB, the measured magnitudes can be normalized using the probe's attenuation.

To capture the magnetic field at different positions around the coil, and to aid in validating the simulation model, the average H-field across the probe was measured at three different positions: center of the coil, corner and edge of the coil turns. Figure **??** shows the test setup as well as the probe positions on the coil.

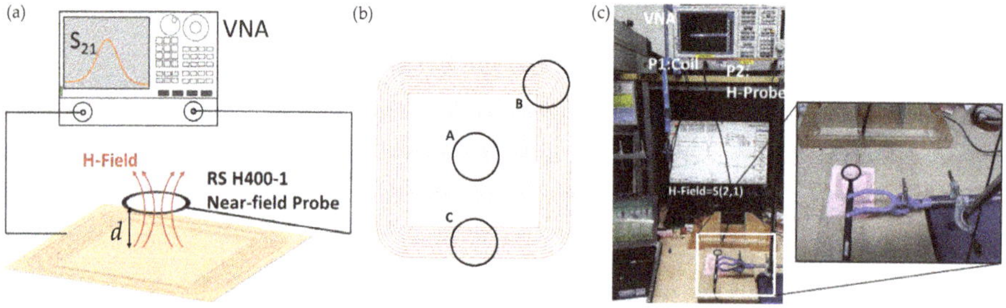

Figure 9. Test setup for measuring the near H-field of the coil: (**a**) schematic, (**b**) probe positions on the coil surface, (**c**) photograph of the setup.

A 3D model of the coils has been built in CST Microwave Studio to observe the coil's near-field and cross validate the measurements. The coil was modelled using a standard lossy copper model and matched using ideal capacitors. A field monitor at 6.78 MHz was used to observe the H-field near the coil, the field at points A, B, and C form Figure **??**. The H-field has been measured as the magnitude of the forward transmission $|S_{21}|$ between the VNA ports, where the normalized H-field was obtained using the maximum H-field measured across the three measurement points. The absolute measured maximum H-field is −10.76 dB compared to −45.5 dB, from the 3D CST simulation model. Figure **??** shows the simulated and measured near-field scans, along with the simulated H-field plot at the surface of the coil.

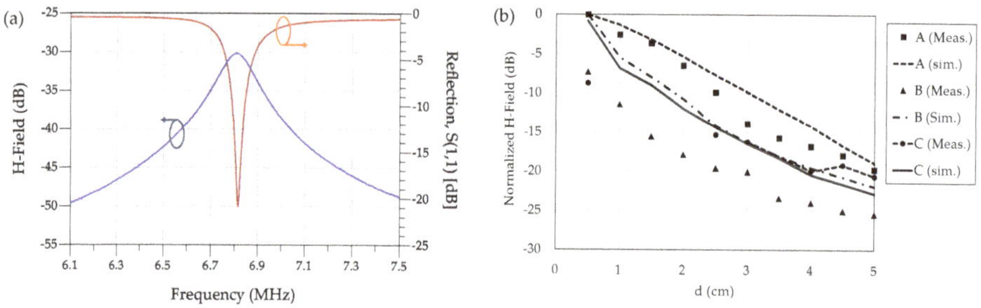

Figure 10. Near-field scans of the coil: (**a**) measured two-port transmission between the coil and the H-field prob, (**b**) simulated and measured normalized H-field at points A, B and C from Figure **??**.

Figure **??** shows the near-field scans obtained from the CST field monitor. It is observed that as with the measurement position, the highest magnetic field at the center of coil (measured position A). Therefore, it is predicted that angular alignment of two WPT coils would have the maximum effect on the WPT efficiency.

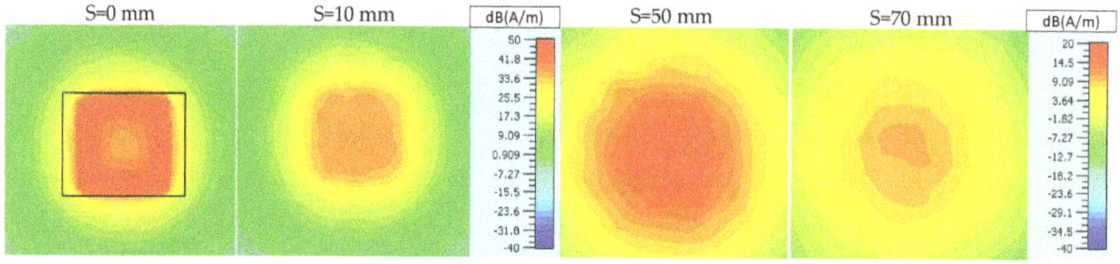

Figure 11. Simulated magnetic near-field at various separations *s* from the surface of the coil.

5.2. Separation-Independent WPT

The forward transmission (S_{21}) of the two coils and the L-matching network (the two-port network shown in Figure ??) has been measured to evaluate the efficiency of the WPT link. The WPT efficiency can be calculated from the forward transmission using $\eta_{WPT} = |S_{21}|^2$. A Rhode and Schwarz ZVB4 VNA, calibrated using a standard Through, Open, Short and Match (TOSM) calibration, was used measure the s-parameters of the network at variable coil separations. Figure ??, shows the coil radiating at 6.78 MHz, showing a 48 KHz ($S_{11} < -10$) impedance bandwidth, at 10 cm separation between the coils, demonstrating uncoupled WPT between the coils.

To demonstrate the separation-independent performance of the two-coil system. The separation between the coils has been studied under two conditions: fully aligned coils (on the vertical XZ plane) and full misalignment (180° misalignment, on the horizontal XY plane); demonstrating the best and worst-case scenarios respectively. Figure ?? shows the test setup of the coils and the different cases considered. The measurements have been performed on multiple 3×3 m^2 planes in a 27 m^3 volume to demonstrate the position-independence. Figure ??, the system's S_{21} is above −17 dB and −28 dB when aligned and misaligned respectively.

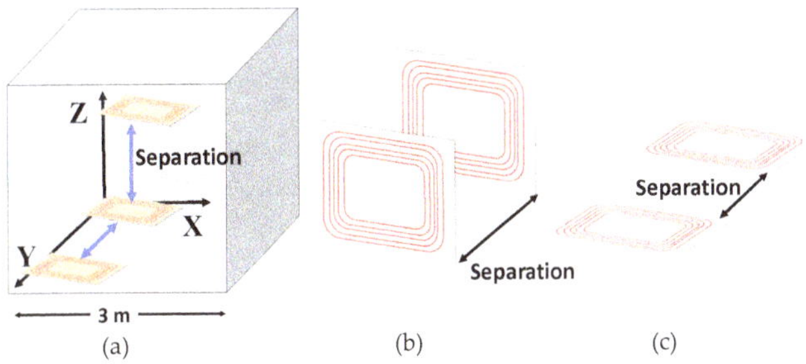

Figure 12. The test conditions of the proposed separation-independent WPT: (**a**) the 3D test volume, (**b**) separation between aligned coils, (**c**) separation between misaligned coils.

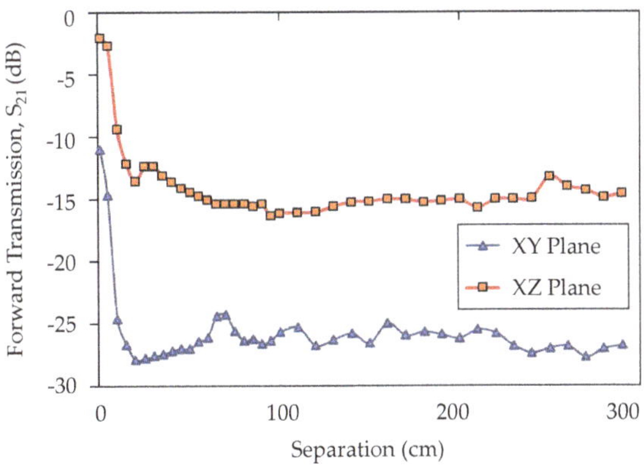

Figure 13. Measured forward transmission between the aligned (XZ plane) and misaligned (XY plane) coils, showing no dependence on the coil separation in the uncoupled region.

While such behavior is predicted for any isotropically radiating system, the position and separation independence achieved is unmatched by any coupled or uncoupled WPT system. In addition, the magnetic field (H-field) radiated in the near-field of the transmitter coil collected more efficiently by the fully aligned receiver, explaining the 10-dB difference in the aligned and misaligned

. Tabesh, M.; Dolatsha, N.; Arbabian, A.; Niknejad, A.M. A Power-Harvesting Pad-Less Millimeter-Sized Radio. *IEEE J. Solid-State Circuits* **2015**, *50*, 962–977. [CrossRef]

. Mizojiri, S.; Shimamura, K. Wireless power transfer via Subterahertz-wave. *Appl. Sci.* **2015**, *8*, 2653. [CrossRef]

. Bao, K.; Zekios, C.L.; Georgakopoulos, S.V. A Wearable WPT System on Flexible Substrates. *IEEE Antennas Wirel. Propag. Lett.* **2019**, *18*, 931–935. [CrossRef]

. Yates, D.; Holmes, A.; Burdett, A. Optimal transmission frequency for ultralow-power short-range radio links. *IEEE Trans. Circuits Syst. I Regul. Pap.* **2004**, *51*, 1405–1413. [CrossRef]

. Seo, D.G.; Ahn, S.H.; Kim, J.H.; Lee, W.S.; Khang, S.T.; Chae, S.C.; Yu, J.W. Power transfer efficiency for distance-adaptive wireless power transfer system. In Proceedings of the 2018 International Applied Computational Electromagnetics Society Symposium (ACES), Denver, CO, USA, 25–29 March 2018.

. Imura, T.; Hori, Y. Maximizing Air Gap and Efficiency of Magnetic Resonant Coupling for Wireless Power Transfer Using Equivalent Circuit and Neumann Formula. *IEEE Trans. Ind. Electron.* **2011**, *58*, 4746–4752. [CrossRef]

. Pinuela, M.; Mitcheson, P.D.; Lucyszyn, S. Ambient RF Energy Harvesting in Urban and Semi-Urban Environments. *IEEE Trans. Microw. Theory Tech.* **2013**, *61*, 2715–2726. [CrossRef]

. Commission, F.C. Federal Communications Commission Washington, D.C. 20554. Available online: https://apps.fcc.gov/edocs_public/attachmatch/DA-09-2425A1.pdf (accessed on 20 January 2020).

. Sun, H.; Geyi, W. Optimum Design of Wireless Power Transmission Systems in Unknown Electromagnetic Environments. *IEEE Access* **2017**, *5*, 2169–3536. [CrossRef]

. Austin, B.A.; Boswell, A.; Perks, M.A. Loss Mechanisms in the Electrically Small Loop Antenna [Antenna Designer's Notebook]. *IEEE Antennas Propag. Mag.* **2014**, *56*, 142–147. [CrossRef]

. Newman, E.; Bohley, P.; Walter, C. Two methods for the measurement of antenna efficiency. *IEEE Trans. Antennas Propag.* **1975**, *23*, 457–461. [CrossRef]

. LaHaie, I. Overview of an Image-Based Technique for Predicting Far-Field Radar Cross Section from Near-Field Measurements.. *IEEE Antennas Propag. Mag.* **2003**, *45*, 159–169. [CrossRef]

energies

MDPI

Communication

Is LiI a Potential Dopant Candidate to Enhance the Thermoelectric Performance in Sb-Free GeTe Systems? A Prelusive Study

Bhuvanesh Srinivasan [1,2,*], David Berthebaud [1] and Takao Mori [2,3,*]

1 CNRS-Saint Gobain-NIMS, UMI 3629, Laboratory for Innovative Key Materials and Structures (LINK), National Institute for Materials Science (NIMS), Tsukuba 305-0044, Japan; David.BERTHEBAUD@cnrs.fr
2 Center for Functional Sensor & Actuator (CFSN) & WPI Center for Materials Nanoarchitechtonics (WPI-MANA), National Institute for Materials Science (NIMS), Tsukuba 305-0044, Japan
3 Graduate School of Pure and Applied Sciences, University of Tsukuba, Tsukuba 305-8671, Japan
* Correspondence: SRINIVASAN.Bhuvanesh@nims.go.jp (B.S.); MORI.Takao@nims.go.jp (T.M.)

Received: 29 November 2019; Accepted: 30 January 2020; Published: 3 February 2020

Abstract: As a workable substitute for toxic PbTe-based thermoelectrics, GeTe-based materials are emanating as reliable alternatives. To assess the suitability of LiI as a dopant in thermoelectric GeTe, a prelusive study of thermoelectric properties of $GeTe_{1-x}LiI_x$ (x = 0–0.02) alloys processed by Spark Plasma Sintering (SPS) are presented in this short communication. A maximum thermoelectric figure of merit, $zT \sim 1.2$, was attained at 773 K for 2 mol% LiI-doped GeTe composition, thanks to the combined benefits of a noted reduction in the thermal conductivity and a marginally improved power factor. The scattering of heat carrying phonons due to the presumable formation of Li-induced "pseudo-vacancies" and nano-precipitates contributed to the conspicuous suppression of lattice thermal conductivity, and consequently boosted the zT of the Sb-free $(GeTe)_{0.98}(LiI)_{0.02}$ sample when compared to that of pristine GeTe and Sb-rich $(GeTe)_x(LiSbTe_2)_2$ compounds that were reported earlier.

Keywords: thermoelectrics; GeTe; LiI as dopant; enhanced power factor; suppressed thermal transport

---

1. Introduction

Thermoelectric (TE) materials and devices have gained limelight due to their ability to reversibly convert waste heat into useful electricity, especially for energy harvesting applications [1–3]. The efficiency of a TE material is quantified by a dimensionless figure of merit, $zT = S^2\sigma T/\kappa$, where σ, S, T and κ are the electrical conductivity, Seebeck coefficient, absolute temperature and total thermal conductivity (which constitute both the electronic part, κ_e, and the lattice part, κ_{latt}), respectively. These transport properties are highly interlinked and there is a greater challenge in decoupling the electrical and thermal transport parameters [4]. To take on these challenges, an array of novel concepts and approaches have been developed and implemented by the thermoelectric community, such as the concepts of band engineering [5,6], introducing nano-scale features [7,8] and discoveries of novel materials [9–13]. Among those state-of-the-art TE materials, the widely studied lead tellurides are limited by their high toxicity for any commercial application, in spite of their high zT [14–18]. Of late, GeTe-based materials have emerged as viable alternatives, as they have proven to exhibit higher TE performance ($zT > 1$), if optimally alloyed/doped with appropriate elements [19]. Such strategies have been effectively adopted on several classes of GeTe-based materials to improve their electrical transport properties (i.e., the power factor, $S^2\sigma$) by the convergence of electronic band valleys [20,21] or by the introduction of resonance states [22,23], and/or to suppress their thermal transport properties by nanostructuring [24]. GeTe undergoes a second-order ferroelectric structural transition from rhombohedral to cubic symmetry at around 700 K [19]. This structural transition holds a key in tuning the TE properties of GeTe. In cubic (c) GeTe, there are two valence bands with a small energy offset, whereas there are several sub-valence bands separated by a large energy offset in the case of rhombohedral (r) GeTe [25]. Therefore, the total number of degenerate valleys in c-GeTe is larger than that of r-GeTe, which will result in a higher total density of states effective mass (m*) without affecting much the mobility, thus resulting in an improved TE performance, as illustrated in the case of In-Sb codoped GeTe [26].

The presented communication tries to explore the suitability of LiI as a dopant for improving the TE performance of GeTe. Our studies on the TE properties of the $(GeTe)_{1-x}(LiI)_x$ system was motivated by a previous report on the *p*-type $(GeTe)_x(LiSbTe_2)_2$ bulk materials, where a strong suppression of κ_{latt} and a slightly improved $zT \sim 1$ at 723 K was reported [27]. It is well established that Sb doping in GeTe helps in improving the band degeneracy and thus ultimately in a better TE performance [19,26]. However a large amount/heavy doping of Sb (~8–10 mol%) is essential for achieving the favorable TE properties [21]. There is no real harm in doping Sb into GeTe, and this strategy has become increasingly incremental although not many reports exists on new strategies to enhance zT in GeTe without employing Sb. Moreover, in almost all cases where Sb was doped into GeTe, there was a severe impact on the electrical conductivity [19,21,26]. Therefore, a critical challenge remains in achieving higher TE performance in the Sb-free GeTe-based systems. This work is one such attempt in exploring the Sb-free GeTe-based thermoelectric compounds. Improvising on that work on $(GeTe)_x(LiSbTe_2)_2$ materials [27], where they attribute the lower thermal transport to the probable "pseudo-vacancy" caused by Li, herein we try to explore the thermoelectric transport properties of Sb-free $(GeTe)_{1-x}(LiI)_x$ solid-state solutions and compare the performance of these Sb-free compounds with that of pristine GeTe and with the previously reported Sb-rich $(GeTe)_x(LiSbTe_2)_2$ system [27].

2. Materials and Methods

The vacuum-sealed tube melt processing route was used to synthesize the samples of $(GeTe)_{1-x}(LiI)_x$ ($x = 0$–0.02). The powders of the crushed ingots were consolidated by Spark Plasma Sintering, SPS (FCT Systeme GmbH), at 773 K for 5 min (holding time) under an axial pressure of approx. 60 MPa. Powder X-ray diffraction patterns were recorded at room temperature in the 2θ range 10–90° with a step size of 0.026° and a scan per step of 400 s using a PANalytical X'Pert Pro diffractometer (Cu K-$L_{2,3}$ radiation, PIXcel 1D detector). The Hall measurements were carried on samples of dimensions ~5 × 5 × 2 mm^3 using a four-point probe setup (Van der Pauw method), where a fixed magnetic field of 0.112 T and DC current of 15 mA was applied. The electrical conductivity and Seebeck coefficients were measured simultaneously from room temperature to 773 K on samples of dimension ~10 × 2 × 2 mm^3 using a commercial instrument (LSR-3, Linseis Inc.) in a He atmosphere. Temperature-dependent thermal diffusivity, D, was measured on disc shaped samples of dimension ~10 mm diameter ×2 mm thickness using the laser flash diffusivity method in a Netzsch LFA-457. The heat capacity, C_p, was derived using the Dulong–Petit relation, $C_p = 3R/M$ (R is the gas constant and M is the molar mass). The total thermal conductivity was obtained from the equation $\kappa = DC_p\rho$, where ρ is the density of the sample measured using Archimedes' principle. The uncertainty in the results for the values of electrical and thermal transport properties was ~5% and ~7%, respectively, and for the overall zT it was ~12% [28]. Error bars are not shown in the figures to increase readability.

3. Results and Discussion

The sharp reflections noticed from the X-ray diffraction (XRD) patterns for $(GeTe)_{1-x}(LiI)_x$ ($x = 0$–0.02) stipulate the phases to be crystalline in nature (Figure 1). The major reflections in both doped and undoped samples could be indexed to the rhombohedral (space group—$R3m$) GeTe phase (PDF# 00-047-1079). The double reflections, i.e., (0 2 4) and (2 2 0), in the range of 2θ values between 41° and 44°, further confirms the presence of the rhombohedral phase. The secondary phase

(Ge-rich) was also present, albeit in a minor proportion, and is in agreement with the previously reported studies [21,23,29,30].

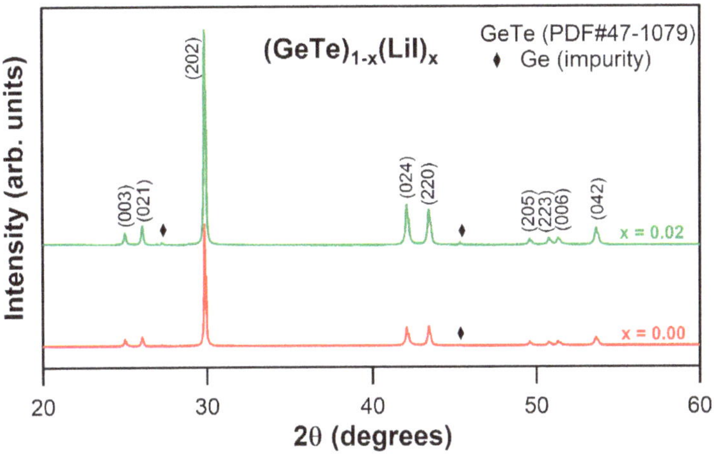

Figure 1. XRD patterns for $(GeTe)_{1-x}(LiI)_x$ (x = 0–0.02) samples.

Results from Hall measurements indicated not much variation in the carrier concentration (n) and mobility (μ), as the pristine GeTe exhibited $n \sim 9.08 \times 10^{20}$ cm^{-3} and $\mu \sim 57$ cm^2 V^{-1} s^{-1}, while the 2 mol% LiI-doped GeTe exhibited $n \sim 9.3 \times 10^{20}$ cm^{-3} and $\mu \sim 54$ cm^2 V^{-1} s^{-1}. As the Hall voltage was positive in the $(GeTe)_{1-x}(LiI)_x$ (x = 0–0.02) system, holes are the dominant charge carriers (p-type). With increasing temperature, the electrical conductivity of the samples slumped (Figure 2a), suggesting a degenerate semi-conducting behavior. The Seebeck coefficient was found to be positive (Figure 2b), thus confirming the p-type behavior in the LiI-doped GeTe, which is in line with the Hall measurement results.

Doping LiI into GeTe marginally enhanced the electrical conductivity, which could probably be due to the LiI-driven higher ionicity and the Seebeck coefficient remained more or less the same as the pristine GeTe. This led to a small but significant improvement in the power factor ($P.F = S^2\sigma$) at elevated temperatures (Figure 2c). In a previously reported Li-Sb codoped study [27], it was postulated that the higher ionicity of Li helped in optimizing the electrical transport properties in those compounds. Considering the close similarities between these material systems, the assumption that LiI-driven ionicity as a cause for improved electrical transport properties in this case seems to hold good. LiI-doping into GeTe markedly decreased the total thermal conductivity (Figure 2d), thanks to the suppressed lattice contribution (Figure 2e). In the case of pristine GeTe, it is well established that the formation energy of the Ge vacancy is quite low, and it is the easily formed intrinsic defect [31,32]. It is reported that Li can fill the cation vacancies and is quite light; hence, it can be regarded as a "pseudo-vacancy", as illustrated by Schröder et al. [27]. Employing Li^+ as a "pseudo-vacancy" might have created a substantial mass difference between the atoms involved, which can potentially disturb the lattice vibrational modes and retard the lattice phononic contribution to the thermal conductivity. Nanostructuring can also be a reason for suppression of lattice contribution, as SPS processing is known to create nano-scale inclusions, defects, dislocations, misfit strains, displacement layers, precipitates and micro/nano-porosity [33–39], all of which, depending on their sizes, are known to scatter phonons of different mean free paths. It can be noted that at around 673 K, there is a change of trend in both electrical and thermal transport properties (for both pristine and LiI-doped GeTe), which is an indication noting the structural transition from the rhombohedral to cubic phase (at ~673 K). LiI-doping in GeTe has helped to simultaneously enhance the power factor and suppress the thermal transport. This cumulative effect of marginally enhanced power factor and significantly suppressed thermal transport has facilitated the LiI-doped GeTe to exhibit a superior thermoelectric figure of merit, zT ~ 1.2 at 773 K (Figure 2f), when compared to the pristine GeTe (zT ~ 0.9 at 773 K). It must be noted that the zT ~ 1.2 at 773 K for the Sb-free $(GeTe)_{0.98}(LiI)_{0.02}$ composition that is reported in this work is considerably higher than what was reported in the literature for the Sb-rich $(GeTe)_x(LiSbTe_2)_2$ composition (zT ~ 1 at 723 K).

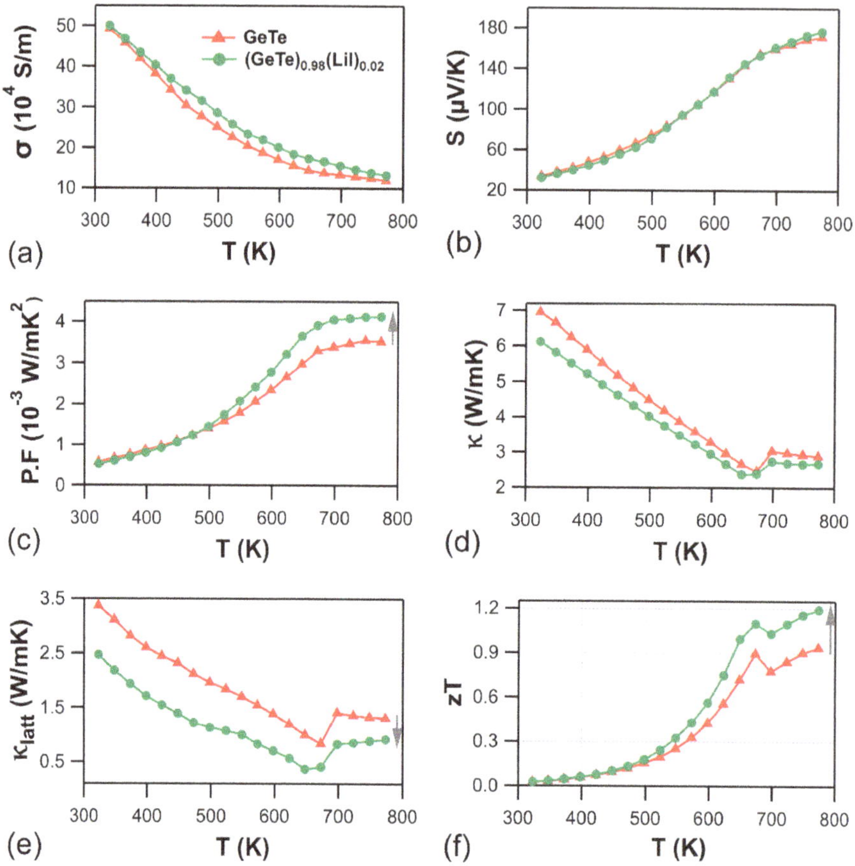

Figure 2. (**a**) Electrical conductivity, σ; (**b**) Seebeck coefficient, S; (**c**) power factor, *P.F*; (**d**) total thermal conductivity, κ; (**e**) lattice contribution to the thermal conductivity, κ_{latt}; and (**f**) thermoelectric figure of merit, zT, for the $(GeTe)_{1-x}(LiI)_x$ (x = 0–0.02) compounds.

As such, this communication has exposed LiI-doped GeTe as a potential candidate for mid-temperature range TE module fabrication, if a suitable n-type counterpart is figured out. More in-depth analysis and experimentation with the variation of LiI content to optimize the dopant concentration (x) in $(GeTe)_{1-x}(LiI)_x$ system are to be explored in the near future to investigate if the zT can be pushed up further (say, closer to ~1.5) with an optimal level of doping and processing conditions.

4. Conclusion

Highly dense samples of $(GeTe)_{1-x}(LiI)_x$ (x = 0–0.02) compounds were successfully synthesized via a two-step process (i.e., melt-processing followed by SPS). Substitution of LiI in GeTe was found to contemporaneously enhance the electrical transport and suppress the thermal transport, thus ultimately boosting up the zT. The Sb-free sample of the $(GeTe)_{0.98}(LiI)_{0.02}$ composition exhibited an improved zT

~ 1.2 at 773 K, which is markedly higher than that of pristine GeTe and the Sb-rich $(GeTe)_x(LiSbTe_2)_2$ compounds that were reported earlier, thus opening up new avenues to explore further and optimize these $(GeTe)_{1-x}(LiI)_x$ compounds for a potential mid-temperature range thermoelectric application.

Author Contributions: B.S. designed and performed the experiments, analyzed the results and wrote the manuscript; project administration, D.B.; funding acquisition, T.M.; all authors contributed to the final manuscript. All authors have read and agreed to the published version of the manuscript.

Funding: This research was initially funded by European Commission's Horizon 2020 research and innovation program under the Marie Skłodowska-Curie grant (GA. 642557, CoACH-ETN), and later by the Japanese Society for the Promotion of Science (JSPS) via postdoctoral fellowship (P19720). Article processing charges were funded by JSPS KAKENHI (19F19720), JST CREST JPMJCR15Q and JPMJCR19Q.

Acknowledgments: Bruno Bureau (Université Rennes 1) and Catherine Boussard-Pledel (CNRS) are thanked for their constant support.

Conflicts of Interest: The authors declare no conflict of interest.

References

1. Mori, T.; Priya, S. Materials for energy harvesting: At the forefront of a new wave. *MRS Bull.* **2018**, *43*, 176–180. [CrossRef]
2. Petsagkourakis, I.; Tybrandt, K.; Crispin, X.; Ohkubo, I.; Satoh, N.; Mori, T. Thermoelectric materials and applications for energy harvesting power generation. *Sci. Technol. Adv. Mater.* **2018**, *19*, 836–862. [CrossRef] [PubMed]
3. Srinivasan, B. Novel Chalcogenide Based Glasses, Ceramics and Polycrystalline Materials for Thermoelectric Application. Ph.D. Thesis, Université Rennes 1, Rennes, France, 2018.
4. Chen, Z.; Jian, Z.; Li, W.; Chang, Y.; Ge, B.; Hanus, R.; Yang, J.; Chen, Y.; Huang, M.; Snyder, G.J.; et al. Lattice Dislocations Enhancing Thermoelectric PbTe in Addition to Band Convergence. *Adv. Mater.* **2017**, *29*, 1606768. [CrossRef] [PubMed]
5. Pei, Y.; Wang, H.; Snyder, G.J. Band Engineering of Thermoelectric Materials. *Adv. Mater.* **2012**, *24*, 6125–6135. [CrossRef] [PubMed]
6. Zhang, Q.; Cao, F.; Liu, W.; Lukas, K.; Yu, B.; Chen, S.; Opeil, C.; Broido, D.; Chen, G.; Ren, Z. Heavy Doping and Band Engineering by Potassium to Improve the Thermoelectric Figure of Merit in p-Type PbTe, PbSe, and $PbTe_{1-y}Se_y$. *J. Am. Chem. Soc.* **2012**, *134*, 10031–10038. [CrossRef] [PubMed]
7. Kanatzidis, M.G. Nanostructured Thermoelectrics: The New Paradigm? *Chem. Mater.* **2010**, *22*, 648–659. [CrossRef]
8. Chen, Z.-G.; Han, G.; Yang, L.; Cheng, L.; Zou, J. Nanostructured thermoelectric materials: Current research and future challenge. *Prog. Nat. Sci. Mater. Int.* **2012**, *22*, 535–549. [CrossRef]
9. Madsen, G.K.H. Automated Search for New Thermoelectric Materials: The Case of LiZnSb. *J. Am. Chem. Soc.* **2006**, *128*, 12140–12146. [CrossRef]
10. Sootsman, J.R.; Chung, D.Y.; Kanatzidis, M.G. New and Old Concepts in Thermoelectric Materials. *Angew. Chem. Int. Ed.* **2009**, *48*, 8616–8639. [CrossRef]
11. Srinivasan, B.; Cui, S.; Prestipino, C.; Gellé, A.; Boussard-Pledel, C.; Ababou-Girard, S.; Trapananti, A.; Bureau, B.; Di Matteo, S. Possible Mechanism for Hole Conductivity in Cu–As–Te Thermoelectric Glasses: A XANES and EXAFS Study. *J. Phys. Chem. C* **2017**, *121*, 14045–14050. [CrossRef]
12. Zhang, T.; Wang, Z.; Srinivasan, B.; Wang, Z.; Zhang, J.; Li, K.; Boussard-Pledel, C.; Troles, J.; Bureau, B.; Wei, L. Ultraflexible Glassy Semiconductor Fibers for Thermal Sensing and Positioning. *ACS Appl. Mater. Interfaces* **2019**, *11*, 2441–2447. [CrossRef] [PubMed]
13. Srinivasan, B.; Boussard-Pledel, C.; Dorcet, V.; Samanta, M.; Biswas, K.; Lefèvre, R.; Gascoin, F.; Cheviré, F.; Tricot, S.; Reece, M.; et al. Thermoelectric Properties of Highly-Crystallized Ge-Te-Se Glasses Doped with Cu/Bi. *Materials* **2017**, *10*, 328. [CrossRef] [PubMed]
14. Snyder, G.J.; Toberer, E.S. Complex thermoelectric materials. *Nat. Mater.* **2008**, *7*, 105–114. [CrossRef] [PubMed]
15. He, J.; Kanatzidis, M.G.; Dravid, V.P. High performance bulk thermoelectrics via a panoscopic approach. *Mater. Today* **2013**, *16*, 166–176. [CrossRef]

16. Zhao, L.-D.; Dravid, V.P.; Kanatzidis, M.G. The panoscopic approach to high performance thermoelectrics. *Energy Environ. Sci.* **2013**, *7*, 251–268. [CrossRef]
17. Srinivasan, B.; Gucci, F.; Boussard-Pledel, C.; Chevirè, F.; Reece, M.J.; Tricot, S.; Calvez, L.; Bureau, B. Enhancement in thermoelectric performance of n-type Pb-deficit Pb-Sb-Te alloys. *J. Alloys Compd.* **2017**, *729*, 198–202. [CrossRef]
18. Srinivasan, B.; Fontaine, B.; Gucci, F.; Dorcet, V.; Saunders, T.G.; Yu, M.; Chevirè, F.; Boussard-Pledel, C.; Halet, J.-F.; Gautier, R.; et al. Effect of the Processing Route on the Thermoelectric Performance of Nanostructured $CuPb_{18}SbTe_{20}$. *Inorg. Chem.* **2018**, *57*, 12976–12986. [CrossRef]
19. Perumal, S.; Roychowdhury, S.; Biswas, K. High performance thermoelectric materials and devices based on GeTe. *J. Mater. Chem. C* **2016**, *4*, 7520–7536. [CrossRef]
20. Roychowdhury, S.; Samanta, M.; Perumal, S.; Biswas, K. Germanium Chalcogenide Thermoelectrics: Electronic Structure Modulation and Low Lattice Thermal Conductivity. *Chem. Mater.* **2018**, *30*, 5799–5813. [CrossRef]
21. Srinivasan, B.; Gellé, A.; Gucci, F.; Boussard-Pledel, C.; Fontaine, B.; Gautier, R.; Halet, J.-F.; Reece, M.J.; Bureau, B. Realizing a stable high thermoelectric zT ~ 2 over a broad temperature range in $Ge_{1-x-y}Ga_xSb_yTe$ via band engineering and hybrid flash-SPS processing. *Inorg. Chem. Front.* **2019**, *6*, 63–73. [CrossRef]
22. Wu, L.; Li, X.; Wang, S.; Zhang, T.; Yang, J.; Zhang, W.; Chen, L.; Yang, J. Resonant level-induced high thermoelectric response in indium-doped GeTe. *NPG Asia Mater.* **2017**, *9*, e343. [CrossRef]
23. Srinivasan, B.; Boussard-Pledel, C.; Bureau, B. Thermoelectric performance of codoped (Bi, In)-GeTe and (Ag, In, Sb)-SnTe materials processed by Spark Plasma Sintering. *Mater. Lett.* **2018**, *230*, 191–194. [CrossRef]
24. Perumal, S.; Roychowdhury, S.; Biswas, K. Reduction of thermal conductivity through nanostructuring enhances the thermoelectric figure of merit in $Ge_{1-x}Bi_xTe$. *Inorg. Chem. Front.* **2016**, *3*, 125–132. [CrossRef]
25. Li, J.; Chen, Z.; Zhang, X.; Sun, Y.; Yang, J.; Pei, Y. Electronic origin of the high thermoelectric performance of GeTe among the p-type group IV monotellurides. *NPG Asia Mater.* **2017**, *9*, e353. [CrossRef]
26. Hong, M.; Chen, Z.-G.; Yang, L.; Zou, Y.-C.; Dargusch, M.S.; Wang, H.; Zou, J. Realizing zT of 2.3 in $Ge_{1-x-y}Sb_xIn_yTe$ via Reducing the Phase-Transition Temperature and Introducing Resonant Energy Doping. *Adv. Mater.* **2018**, *30*, 1705942. [CrossRef]
27. Schröder, T.; Schwarzmüller, S.; Stiewe, C.; de Boor, J.; Hölzel, M.; Oeckler, O. The Solid Solution Series $(GeTe)_x(LiSbTe_2)_2$ ($1 \leq x \leq 11$) and the Thermoelectric Properties of $(GeTe)_{11}(LiSbTe_2)_2$. *Inorg. Chem.* **2013**, *52*, 11288–11294. [CrossRef]
28. Simon, J.; Guélou, G.; Srinivasan, B.; Berthebaud, D.; Mori, T.; Maignan, A. Exploring the thermoelectric behavior of spark plasma sintered $Fe_{7-x}Co_xS_8$ compounds. *J. Alloys Compd.* **2020**, *819*, 152999. [CrossRef]
29. Srinivasan, B.; Gautier, R.; Gucci, F.; Fontaine, B.; Halet, J.-F.; Chevirè, F.; Boussard-Pledel, C.; Reece, M.J.; Bureau, B. Impact of Coinage Metal Insertion on the Thermoelectric Properties of GeTe Solid-State Solutions. *J. Phys. Chem. C* **2018**, *122*, 227–235. [CrossRef]
30. Srinivasan, B.; Gellé, A.; Halet, J.-F.; Boussard-Pledel, C.; Bureau, B. Detrimental Effects of Doping Al and Ba on the Thermoelectric Performance of GeTe. *Materials* **2018**, *11*, 2237. [CrossRef]
31. Levin, E.M. Effects of Ge substitution in GeTe by Ag or Sb on the Seebeck coefficient and carrier concentration derived from ^{125}Te NMR. *Phys. Rev. B* **2016**, *93*, 045209. [CrossRef]
32. Edwards, A.H.; Pineda, A.C.; Schultz, P.A.; Martin, M.G.; Thompson, A.P.; Hjalmarson, H.P.; Umrigar, C.J. Electronic structure of intrinsic defects in crystalline germanium telluride. *Phys. Rev. B* **2006**, *73*, 045210. [CrossRef]
33. Hong, M.; Wang, Y.; Feng, T.; Sun, Q.; Xu, S.; Matsumura, S.; Pantelides, S.T.; Zou, J.; Chen, Z.-G. Strong Phonon–Phonon Interactions Securing Extraordinary Thermoelectric $Ge_{1-x}Sb_xTe$ with Zn-Alloying-Induced Band Alignment. *J. Am. Chem. Soc.* **2019**, *141*, 1742–1748. [CrossRef] [PubMed]
34. Zheng, Z.; Su, X.; Deng, R.; Stoumpos, C.; Xie, H.; Liu, W.; Yan, Y.; Hao, S.; Uher, C.; Wolverton, C.; et al. Rhombohedral to Cubic Conversion of GeTe via MnTe Alloying Leads to Ultralow Thermal Conductivity, Electronic Band Convergence, and High Thermoelectric Performance. *J. Am. Chem. Soc.* **2018**, *140*, 2673–2686. [CrossRef] [PubMed]
35. Perumal, S.; Samanta, M.; Ghosh, T.; Shenoy, U.S.; Bohra, A.K.; Bhattacharya, S.; Singh, A.; Waghmare, U.V.; Biswas, K. Realization of High Thermoelectric Figure of Merit in GeTe by Complementary Co-doping of Bi and In. *Joule* **2019**, *3*, 2565–2580. [CrossRef]

36. Bayikadi, K.S.; Sankar, R.; Wu, C.T.; Xia, C.; Chen, Y.; Chen, L.-C.; Chen, K.-H.; Chou, F.-C. Enhanced thermoelectric performance of GeTe through in situ microdomain and Ge-vacancy control. *J. Mater. Chem. A* **2019**, *7*, 15181–15189. [CrossRef]
37. Biswas, K.; He, J.; Blum, I.D.; Wu, C.-I.; Hogan, T.P.; Seidman, D.N.; Dravid, V.P.; Kanatzidis, M.G. High-performance bulk thermoelectrics with all-scale hierarchical architectures. *Nature* **2012**, *489*, 414–418. [CrossRef]
38. Grasso, S.; Kim, E.-Y.; Saunders, T.; Yu, M.; Tudball, A.; Choi, S.-H.; Reece, M. Ultra-Rapid Crystal Growth of Textured SiC Using Flash Spark Plasma Sintering Route. *Cryst. Growth Design* **2016**, *16*, 2317–2321. [CrossRef]
39. Qian Zhang, Y.L. Increased thermoelectric performance by Cl doping in nanostructured $AgPb_{18}SbSe_{20-x}Cl_x$. *Nano Energy* **2013**, *2*, 1121–1127. [CrossRef]

energies

MDPI

Article

Novel Flexible Triboelectric Nanogenerator based on Metallized Porous PDMS and Parylene C

Massimo Mariello [1,2,*], Elisa Scarpa [1], Luciana Algieri [3], Francesco Guido [1], Vincenzo Mariano Mastronardi [1], Antonio Qualtieri [1] and Massimo De Vittorio [1,2]

1 Istituto Italiano di Tecnologia, Center for Biomolecular Nanotechnologies, 73010 Arnesano (Lecce), Italy; elisa.scarpa@iit.it (E.S.); francesco.guido@iit.it (F.G.); vincenzo.mastronardi@iit.it (V.M.M.); antonio.qualtieri@iit.it (A.Q.); massimo.devittorio@iit.it (M.D.V.)
2 Dipartimento di Ingegneria dell'Innovazione, Università del Salento, 73100 Lecce, Italy
3 Piezoskin s.r.l., 73010 Arnesano (Lecce), Italy; lalgieri@piezoskin.com
* Correspondence: massimo.mariello@iit.it; Tel.: +39-0832-1816-200

Received: 9 March 2020; Accepted: 26 March 2020; Published: 2 April 2020

Abstract: Triboelectric nanogenerators (TENGs) have recently become a powerful technology for energy harvesting and self-powered sensor networks. One of their main advantages is the possibility to employ a wide range of materials, especially for fabricating inexpensive and easy-to-use devices. This paper reports the fabrication and preliminary characterization of a novel flexible triboelectric nanogenerator which could be employed for driving future low power consumption wearable devices. The proposed TENG is a single-electrode device operating in contact-separation mode for applications in low-frequency energy harvesting from intermittent tapping loads involving the human body, such as finger or hand tapping. The novelty of the device lies in the choice of materials: it is based on a combination of a polysiloxane elastomer and a poly (para-xylylene). In particular, the TENG is composed, sequentially, of a poly (dimethylsiloxane) (PDMS) substrate which was made porous and rough with a steam-curing step; then, a metallization layer with titanium and gold, deposited on the PDMS surface with an optimal substrate–electrode adhesion. Finally, the metallized structure was coated with a thin film of parylene C serving as friction layer. This material provides excellent conformability and high charge-retaining capability, playing a crucial role in the triboelectric process; it also makes the device suitable for employment in harsh, wet environments owing to its inertness and barrier properties. Preliminary performance tests were conducted by measuring the open-circuit voltage and power density under finger tapping (~2 N) at ~5 Hz. The device exhibited a peak-to-peak voltage of 1.6 V and power density peak of 2.24 mW/m^2 at ~0.4 MΩ. The proposed TENG demonstrated ease of process, simplicity, cost-effectiveness, and flexibility.

Keywords: triboelectric nanogenerator (TENG); mechanical energy harvesting; single-electrode; tapping; flexibility; porous/rough PDMS; parylene C

1. Introduction

Flexible, wearable electronics have attracted plenty of research and technology attention owing to their advantageous properties over rigid electronic systems and devices of softness, lightweight, comfort, and adaptability [1–4]. Combining these features with functional materials and specific architectures has led to the employment of flexible devices for several applications, e.g., physiological monitoring and chemical sensing [5–9], soft robotics [10–12], energy harvesting [13–17], and motion detection [18–20]. The recently increasing importance of self-powered electronics has led to the development of integrated systems with energy harvesting technologies.

However, the demand for thin, compliant, and lightweight systems for replacing batteries and supplying microelectronic devices has led to plenty of research efforts to devise efficient, miniaturized power sources. Conventional batteries, in fact, are characterized by limited capacity, necessity of periodic charging, difficulty of miniaturization, and harmfulness of employed material; thus, new energy supplying systems have been studied and proposed in recent years. Mechanical energy harvesting from ambient vibrations or friction is a very promising field and several mechanisms to convert mechanical into electrical energy (energy harvesters) have been exploited in the last years, e.g., electromagnetic [21–24], piezoelectric [25–30], and triboelectric [31–35]. Among them, the triboelectric principle is an attractive option [36]. The triboelectric effect is an electrification process induced by the contact between two different materials: the friction causes a charge separation depending on the different polarity of the materials in contact [37,38].

Triboelectric nanogenerator (TENG) was first implemented by Prof. Z. L. Wang's group in 2012 for the conversion of small-scale mechanical energy into electrical energy. Since then, this technology has spread and improved with the advent of new device architectures and the employment of new functional materials. TENG devices exhibit several advantages compared to other types of energy harvesting or sensing devices: light mass, low density and cost, they are not based on magnets or coils like classical electromagnetic generators, and they exhibit their best performances at low frequencies ($<$10 Hz). Thus, they are suitable for applications regarding human body motion and ocean waves [31,37,39–41].

TENG technology exploits four main basic operation modes, as previously reported in [37]: (i) vertical contact-separation mode; (ii) lateral sliding mode; (iii) single-electrode mode; (iv) freestanding triboelectric-layer mode. Although the single-electrode devices exhibit lower ranges of generated voltages and powers [42], they are generally characterized by simpler architectures, ease of process, and suitability of energy generation by intermittent tapping onto the device. For these reasons, they are suitable for applications involving the human body, such as finger and hand tapping or walking on the floor [37,43].

Many examples of triboelectric energy harvesters have been proposed worldwide, with different operation modes [31,33,39,44–50]. In this respect, the search and exploitation of new friction materials is ongoing [32,36,39,51–53], as well as new techniques for enhancing triboelectric charge transfer based on surface treatments, nanostructured patterns, or modifications of roughness and porosity [43,52,54,55].

Zhang et al. [56] demonstrated a flexible single-electrode TENG based on double-sided PDMS surface nanostructures (nanopillars with diameter of about 200 nm, made by imprint lithography): the device has good output performance at low frequencies, up to 160 V, 3 μA, and 423.8 mW/m^2 for the open-circuit voltage, short-circuit current, and power density, respectively, under applied pressure forces in the range of 0.5–50 N.

Nguyen et al. [57] showed the possibility to enhance the output performances of a peptide-based piezoelectric nanogenerator by utilizing a single-electrode TENG without additional conductive electrodes based on the vertical contact-separation of two triboelectric materials, kapton and poly (ethyleneterephtalate) (PET). The open-circuit voltage and short-circuit current of the triboelectric component under an applied force of 50 N are about 1.6 V and 20 nA, respectively, which can be superimposed to the outputs of the piezoelectric part, 0.7 V and 30 nA, respectively.

Wang et al. [58] reported the fabrication of a single-electrode TENG based on sponge-like porous poly (tetrafluoroethylene) (PTFE) thin films made by using deionized water as soft template. In this case, porosity contributes to enhancing the output performances by a factor of 1.8 on the generated voltage compared to the non-porous control; an output voltage of 1.1 V can be achieved by pressing the device with a bare human hand.

The need and demand for simple fabrication techniques for TENGs is ongoing, and major efforts for developing or employing flexible cost-effective materials are still required.

This paper reports on the low-cost fabrication of a novel flexible single-electrode TENG operating in contact-separation mode based on metallized porous PDMS and parylene C (pC). The novelty of the device lies in the choice of materials: the porosity of metallized PDMS and the charge-retaining capability of pC play a crucial role in the triboelectric working mechanism.

Parylene C was chosen as friction layer by virtue of its desirable outstanding properties of optimal conformality, chemical inertness, and compactness: it is an excellent coating for different types of substrates and the synthesis process is very efficient in allowing complete control of the deposition parameters [59]. Parylene is used as a protective coating for biomedical devices and microelectronics [60], but it also reveals great potential as a charge accumulation layer during the triboelectrification process because surface charge densities higher than 2.0 mC/m^2 may be achieved [54,61].

2. Materials and Methods

2.1. Materials

Silicon wafers or glass slides were used as substrates for the deposition of PDMS by spin-coating.

PDMS (Sylgard 184 Silicone Elastomer) was supplied by Dow Corning Corporation in two forms: a viscous uncured pre-polymer and a curing agent. Parylene C was provided by Specialty Coating Systems in form of dimer powders. Kapton HN 25 μm-thick foils were supplied by DuPont.

2.2. Fabrication of Flexible TENG

In Figure 1A,B a schematic representation and captured images of the fabricated devices are reported, showing an active area of the device of 2 × 2 cm^2, while Figure 1C illustrates the steps of the fabrication process.

After curing (100 °C, 10 min) a 160 μm-thick spin-coated PDMS film (Figure 1C(i)), and a thinner PDMS layer (15–20 μm) were deposited at a higher spinning speed (2000 rpm) and cured with water vapor (100 °C) for 5 min (Figure 1C(ii)).

A flask filled with distilled water was placed on a hot plate (at 120 °C) and connected through a temperature-resistant rubber tube to a glass chamber containing the samples (see Figure 2A). The tube inlet was controlled by a valve which was opened when the water inside the flasks reached a constant temperature and pressure. The tube outlet was enlarged with a glass funnel in order to achieve a wide vapor jet. The steam-curing step, as previously described in [62], was adopted to produce asperities and pores which allow the adhesion of metal layers onto PDMS and favor the charge generation during contact of friction layers in the triboelectric device. The residual water droplets that condensed on the PDMS surface were removed under a nitrogen flow and by heating the sample (120 °C, 10 min).

The metal electrodes of the devices were made by metallizing the PDMS substrate by physical vapor deposition processes: titanium (50 nm) was first deposited as adhesion promoter (Figure 1C(iii)) by reactive sputtering and then gold (50 nm) was thermally evaporated as exposed electrode (Figure 1C(iv]). After peeling off the flexible patch (Figure 1C(v)), it was attached onto a 25 μm-thick kapton foil and used as support for the connections (Figure 1C(vi)): an electrical wire was connected using a rivet and a ~2.5 μm-thick parylene C film was then deposited by room temperature chemical vapor deposition (RT-CVD) step (Figure 1C(vii)).

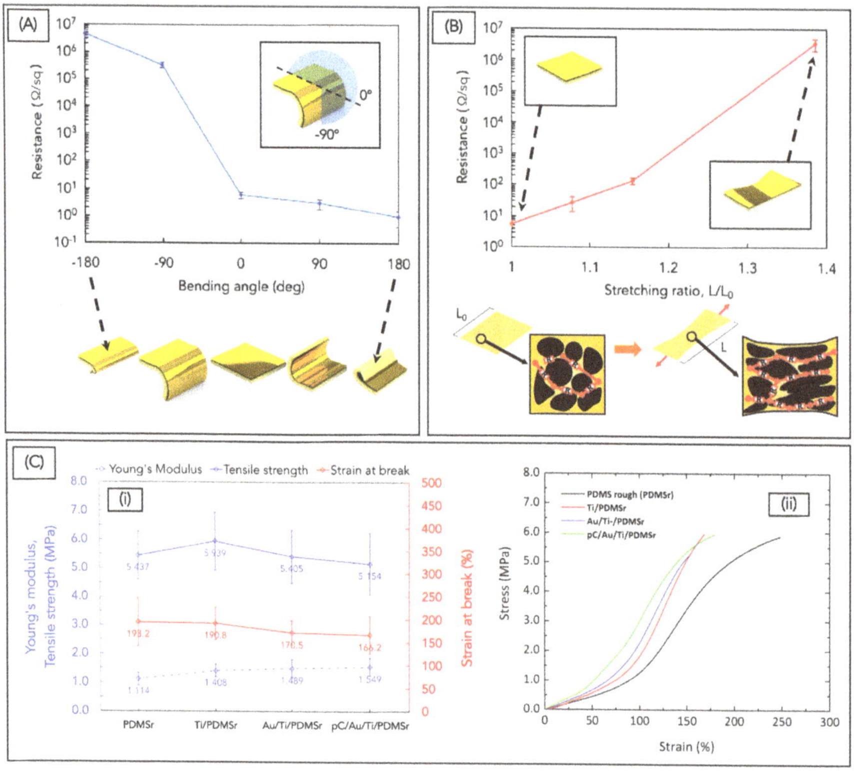

Figure 3. (**A**) Measurement of electrical resistance of the metallized PDMS patch as a function of the bending angle. (**B**) Measurement of electrical resistance of the metallized PDMS patch as a function of the stretching ratio; the insets show the conductive pathways inside the porous morphology of the patch. (**C**) Dynamic mechanical analysis (DMA) of the PDMS rough patch with the overlying layers. (i) Summary plot of the Young's modulus, tensile strength, and strain at break of neat rough PDMS (PDMSr) and covered with Ti, Au/Ti, and pC/Au/Ti. (ii) Stress–strain curves obtained after the tensile tests of the same samples.

Stress–strain curves obtained by dynamic mechanical analysis are reported in Figure 3C(ii) for the rough PDMS patch (PDMSr) and the patches metallized with Ti, Au/Ti, and covered with a parylene layer. The Young's modulus, tensile strength, and strain at break of the samples were deduced from the curves for each sample and summarized in Figure 3C(i): the range for the measured quantities is comparable for the differently metallized patches and a slight increase in the Young's modulus can be observed, due to the deposition of metal and parylene films onto the PDMS patch. The PDMS-based triboelectric device without the kapton support exhibits a Young's modulus of 1.549 MPa, a tensile strength of 5.154 MPa, and a strain at break of 166.2%.

The presented TENG was fabricated as single-electrode device for energy harvesting from a low-frequency source, in particular, from human body motion. Performance preliminary tests were thus conducted by measuring the open-circuit voltage under finger tapping. The average applied force and frequency of the load were measured with a force resistive sensor (FSR 402, Interlink Electronics Sensor Technologies), resulting as ~2 N and ~5 Hz. The output voltage was measured with the oscilloscope for different parylene (pC) thicknesses: as illustrated by the plot in Figure 4C, the voltage decreases with increasing pC thickness, mainly due to the decrease in the capacity of the triboelectric device as well as the higher rigidity of the layered structure. In particular, the device exhibited a peak-to-peak voltage of ~1.6 V for 1 µm-thick pC, and up to ~0.6 V for 6 µm-thick pC. Consequently, the generation tests were performed by choosing a pC thickness of 1 µm: the maximum peak-to-peak open-circuit voltage and short-circuit current, measured with oscilloscope and source meter respectively, were ~1.6 V and ~0.15 µA (Figure 4A,B). The power density curve is reported in Figure 4D, as calculated after measuring the average peak-to-peak voltage drop on the resistive loads connected to the TENG, according to the following expression:

$$p = \frac{\langle P \rangle}{S} = \frac{V^2/R_\Omega}{S} \quad (1)$$

where p is the output power density (mW/m^2) $\langle P \rangle$ is the output power (mW); S is the TENG active area (m^2); R_Ω, V are the resistance (Ω) and the peak-to-peak voltage drop (V) on the resistive load, respectively. The maximum power is reached when the device is closed on an optimal load matching the device electrical impedance [63]: the peak of power density was 2.24 mW/m^2 at ~0.4 MΩ (Figure 4D).

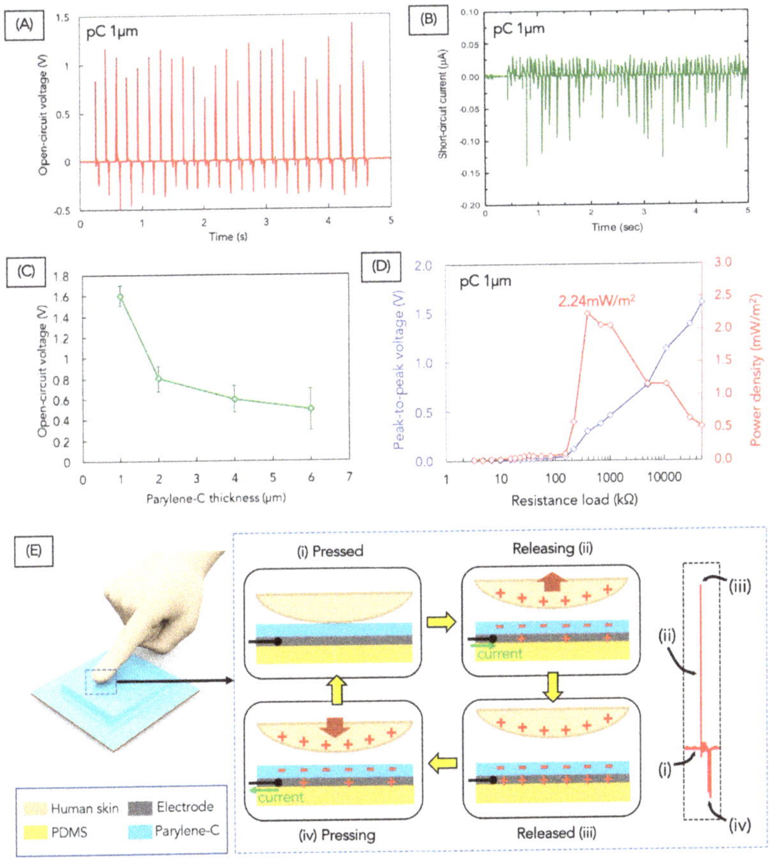

Figure 4. Open-circuit voltage (**A**) and short-circuit current (**B**) generated by the proposed triboelectric nanogenerator (TENG) with the triboelectric parylene C layer of 1 µm, under finger tapping of ~2N at ~5Hz. (**C**) Open-circuit voltage as a function of the parylene C thickness. (**D**) Peak-to-peak voltage and power density generated by the TENG with parylene C thickness of 1 µm as a function of the resistive load. (**E**) Proposed working principle for the TENG under finger tapping: the steps of the press–release cycle are illustrated and indicated for the inset of the output voltage signal.

In Figure 4E, the proposed working mechanism of the TENG is illustrated. The exact operation mode is defined as conductor-to-dielectric single-electrode vertical contact-separation TENG, and the conductor approaching the device is represented by the naked finger. At the initial state of the press–release cycle (Figure 4E(i)), the finger and the upper layer of parylene C are brought into contact, such that surface charge transfer takes place due to the friction and contact electrification effect. According to the triboelectric series, human dry skin acts as a positive tribo-material [42,64,65] so positive charges are generated on the skin surface and the same amount of negative charges form on the surface of parylene coating according to the charge conservation principle. Therefore, positive charges are electrostatically induced on the primary electrode underneath parylene (Ti/Au metallization). The triboelectric charges on parylene are confined on the surface because it is an insulator, so during contact (Figure 4E(i)), charges with opposite signs reside almost on the same plane, and no electrical potential difference is established between the two sides. When the finger starts to separate from the parylene surface (Figure 4E(ii)), a potential difference arises between the primary electrode and the ground under open-circuit conditions, increasing with the vertical finger–parylene separation distance (up to 1–2 cm for complete release in Figure 4E(iii)). In short-circuit conditions, the voltage drop drives an electron transfer, thus generating an instantaneous current from the ground to the electrode, forcing them to have the same potential. A successive approach of the finger to parylene (Figure 4E(iv)) triggers the same mechanisms, but in the opposite way, because the potential difference decreases and the electrons are driven from the primary electrode to the ground.

The TENG proposed in the present work is envisioned to be easily employed in distributed arrays to meet the power requirements of various wearable, flexible energy harvesting devices. Moreover, the adoption of parylene C as an external encapsulating friction layer makes the device suitable for employment in harsh environments; for instance, for scavenging energy from falling raindrops, since the protective anti-corrosion and anti-biofouling behavior of parylene have been demonstrated in previous works [60,66–69].

4. Conclusions

Our research shows an easy fabrication method for a flexible contact-separation mode triboelectric nanogenerator (TENG) based on novel combinations of materials, in essence, a porous PDMS substrate metallized with titanium and gold and coated with a conformal thin film of parylene C.

The TENG is a single-electrode device, so it is characterized by simpler architecture and ease of process if compared to more complex operation modes. It is particularly suitable for energy generation by intermittent tapping onto the device and, thus, for applications involving the human body such as finger and hand tapping or walking on the floor.

Performance preliminary tests were conducted by measuring the open-circuit voltage and power density under finger tapping (~2 N) at ~5 Hz. The device exhibited a peak-to-peak voltage of ~1.6 V and the peak of power density was 2.24 mW/m^2 at ~0.4 MΩ. The working mechanism of the TENG was proposed for contact with human skin based on the experimental results. The use of parylene C lets us envision the possibility to employ the device in harsh environments, where humidity or other aggressive agents are present.

Author Contributions: M.M. conceived and performed the experiments; E.S. contributed for the realization of the setup for steam-curing of PDMS; L.A., F.G. and V.M.M. contributed tools for electrical characterization tests; A.Q. contributed for observation of the PDMS surface morphology with SEM; M.M analyzed the data and wrote the paper; M.D.V. supervised the experiments. All authors have read and agreed to the published version of the manuscript.

Funding: The APC was funded by the following projects: "HORIZON 2020" PON I&C 2014-2020 "Recupero di Energia Meccanica da fluidi per internet delle cose e sensori remote (REM)" n.F/050111/01-03/X32; PON "R&I" 2014-2020 "SE4I – Smart Energy Efficiency & Environment for Industry" n. ARS01_01137.

Conflicts of Interest: The authors declare no conflicts of interest.

References

1. Chen, H.; Song, Y.; Guo, H.; Miao, L.; Chen, X.; Su, Z.; Zhang, H. Hybrid porous micro structured finger skin inspired self-powered electronic skin system for pressure sensing and sliding detection. *Nano Energy* **2018**, *51*, 496–503. [CrossRef]
2. Wang, S.; Xu, J.; Wang, W.; Wang, G.-J.N.; Rastak, R.; Molina-Lopez, F.; Chung, J.W.; Niu, S.; Feig, V.R.; Lopez, J.; et al. Skin electronics from scalable fabrication of an intrinsically stretchable transistor array. *Nature* **2018**, *555*, 83–88. [CrossRef] [PubMed]
3. Chortos, A.; Liu, J.; Bao, Z. Pursuing prosthetic electronic skin. *Nat. Mater.* **2016**, *15*, 937–950. [CrossRef] [PubMed]
4. Gu, Y.; Zhang, T.; Chen, H.; Wang, F.; Pu, Y.; Gao, C.; Li, S. Mini Review on Flexible and Wearable Electronics for Monitoring Human Health Information. *Nanoscale Res. Lett.* **2019**, *14*, 263. [CrossRef] [PubMed]
5. Wang, C.; Li, X.; Hu, H.; Zhang, L.; Huang, Z.; Lin, M.; Zhang, Z.; Yin, Z.; Huang, B.; Gong, H.; et al. Monitoring of the central blood pressure waveform via a conformal ultrasonic device. *Nat. Biomed. Eng.* **2018**, *2*, 687–695. [CrossRef]
6. Gao, W.; Emaminejad, S.; Nyein, H.Y.Y.; Challa, S.; Chen, K.; Peck, A.; Fahad, H.M.; Ota, H.; Shiraki, H.; Kiriya, D.; et al. Fully integrated wearable sensor arrays for multiplexed in situ perspiration analysis. *Nature* **2016**, *529*, 509–514. [CrossRef]
7. Chung, M.; Fortunato, G.; Radacsi, N. Wearable flexible sweat sensors for healthcare monitoring: A review. *J. R. Soc. Interface* **2019**, *16*, 20190217. [CrossRef]
8. Tee, B.C.-K.; Chortos, A.; Berndt, A.; Nguyen, A.K.; Tom, A.; McGuire, A.; Lin, Z.C.; Tien, K.; Bae, W.-G.; Wang, H.; et al. A skin-inspired organic digital mechanoreceptor. *Science* **2015**, *350*, 313–316. [CrossRef]
9. Jang, K.-I.; Li, K.; Chung, H.U.; Xu, S.; Jung, H.N.; Yang, Y.; Kwak, J.W.; Jung, H.H.; Song, J.; Yang, C.; et al. Self-assembled three dimensional network designs for soft electronics. *Nat. Commun.* **2017**, *8*, 15894. [CrossRef]
10. Byun, J.; Lee, Y.; Yoon, J.; Lee, B.; Oh, E.; Chung, S.; Lee, T.; Cho, K.-J.; Kim, J.; Hong, Y. Electronic skins for soft, compact, reversible assembly of wirelessly activated fully soft robots. *Sci. Robot.* **2018**, *3*. [CrossRef]
11. Lai, Y.-C.; Deng, J.; Liu, R.; Hsiao, Y.-C.; Zhang, S.L.; Peng, W.; Wu, H.-M.; Wang, X.; Wang, Z.L. Actively Perceiving and Responsive Soft Robots Enabled by Self-Powered, Highly Extensible, and Highly Sensitive Triboelectric Proximity- and Pressure-Sensing Skins. *Adv. Mater.* **2018**, *30*, e1801114. [CrossRef] [PubMed]
12. Ju, H.; Jeong, J.; Kwak, P.; Kwon, M.; Lee, J. Robotic Flexible Electronics with Self-Bendable Films. *Soft Robot.* **2018**, *5*, 710–717. [CrossRef] [PubMed]
13. Park, S.; Heo, S.W.; Lee, W.; Inoue, D.; Jiang, Z.; Yu, K.; Jinno, H.; Hashizume, D.; Sekino, M.; Yokota, T.; et al. Self-powered ultra-flexible electronics via nano-grating-patterned organic photovoltaics. *Nature* **2018**, *561*, 516–521. [CrossRef] [PubMed]
14. Chen, J.; Huang, Y.; Zhang, N.; Zou, H.; Liu, R.; Tao, C.; Fan, X.; Wang, Z.L. Micro-cable structured textile for simultaneously harvesting solar and mechanical energy. *Nat. Energy* **2016**, *1*, 16138. [CrossRef]
15. Kim, S.J.; Lee, H.E.; Choi, H.; Kim, Y.; We, J.H.; Shin, J.S.; Lee, K.J.; Cho, B.J. High-Performance Flexible Thermoelectric Power Generator Using Laser Multiscanning Lift-Off Process. *ACS Nano* **2016**, *10*, 10851–10857. [CrossRef]
16. García Núñez, C.; Manjakkal, L.; Dahiya, R. Energy autonomous electronic skin. *Npj Flex. Electron.* **2019**, *3*, 1–24. [CrossRef]
17. Dagdeviren, C.; Yang, B.D.; Su, Y.; Tran, P.L.; Joe, P.; Anderson, E.; Xia, J.; Doraiswamy, V.; Dehdashti, B.; Feng, X.; et al. Conformal piezoelectric energy harvesting and storage from motions of the heart, lung, and diaphragm. *Proc. Natl. Acad. Sci. USA* **2014**, *111*, 1927–1932. [CrossRef]
18. Lee, S.; Reuveny, A.; Reeder, J.; Lee, S.; Jin, H.; Liu, Q.; Yokota, T.; Sekitani, T.; Isoyama, T.; Abe, Y.; et al. A transparent bending-insensitive pressure sensor. *Nat. Nanotechnol.* **2016**, *11*, 472–478. [CrossRef]
19. Guo, S.-Z.; Qiu, K.; Meng, F.; Park, S.H.; McAlpine, M.C. 3D Printed Stretchable Tactile Sensors. *Adv. Mater.* **2017**, *29*, 1701218. [CrossRef]
20. Liu, C.; Huang, N.; Xu, F.; Tong, J.; Chen, Z.; Gui, X.; Fu, Y.; Lao, C. 3D Printing Technologies for Flexible Tactile Sensors toward Wearable Electronics and Electronic Skin. *Polymers* **2018**, *10*, 629. [CrossRef]

21. Liu, H.; Hou, C.; Lin, J.; Li, Y.; Shi, Q.; Chen, T.; Sun, L.; Lee, C. A non-resonant rotational electromagnetic energy harvester for low-frequency and irregular human motion. *Appl. Phys. Lett.* **2018**, *113*, 203901. [CrossRef]
22. Liu, S.; Li, P.; Yang, Y. On the design of an electromagnetic aeroelastic energy harvester from nonlinear flutter. *Meccanica* **2018**, *53*, 2807–2831. [CrossRef]
23. Sardini, E.; Serpelloni, M. Nonlinear electromagnetic generators with polymeric materials for power harvesting from vibrations. *Procedia Eng.* **2010**, *5*, 1168–1171. [CrossRef]
24. Wang, D.-A.; Chiu, C.-Y.; Pham, H.-T. Electromagnetic energy harvesting from vibrations induced by Kármán vortex street. *Mechatronics* **2012**, *22*, 746–756. [CrossRef]
25. Akaydin, H.D.; Elvin, N.; Andreopoulos, Y. Energy Harvesting from Highly Unsteady Fluid Flows using Piezoelectric Materials. *J. Intell. Mater. Syst. Struct.* **2010**, *21*, 1263–1278. [CrossRef]
26. Dagdeviren, C.; Joe, P.; Tuzman, O.L.; Park, K.-I.; Lee, K.J.; Shi, Y.; Huang, Y.; Rogers, J.A. Recent progress in flexible and stretchable piezoelectric devices for mechanical energy harvesting, sensing and actuation. *Extreme Mech. Lett.* **2016**, *9*, 269–281. [CrossRef]
27. Hobeck, J.; Inman, D. Artificial piezoelectric grass for energy harvesting from turbulence-induced vibration. *Smart Mater. Struct.* **2012**, *21*, 105024. [CrossRef]
28. Fan, K.; Liu, Z.; Liu, H.; Wang, L.; Zhu, Y.; Yu, B. Scavenging energy from human walking through a shoe-mounted piezoelectric harvester. *Appl. Phys. Lett.* **2017**, *110*, 143902. [CrossRef]
29. Wong, V.-K.; Ho, J.-H.; Chai, A.-B. Performance of a piezoelectric energy harvester in actual rain. *Energy* **2017**, *124*, 364–371. [CrossRef]
30. Petrini, F.; Gkoumas, K. Piezoelectric energy harvesting from vortex shedding and galloping induced vibrations inside HVAC ducts. *Energy Build.* **2018**, *158*, 371–383. [CrossRef]
31. Khan, U.; Kim, S.-W. Triboelectric Nanogenerators for Blue Energy Harvesting. *ACS Nano* **2016**, *10*, 6429–6432. [CrossRef] [PubMed]
32. Choi, A.Y.; Lee, C.J.; Park, J.; Kim, D.; Kim, Y.T. Corrugated Textile based Triboelectric Generator for Wearable Energy Harvesting. *Sci. Rep.* **2017**, *7*, 45583. [CrossRef] [PubMed]
33. Hou, T.-C.; Yang, Y.; Zhang, H.; Chen, J.; Chen, L.-J.; Lin Wang, Z. Triboelectric nanogenerator built inside shoe insole for harvesting walking energy. *Nano Energy* **2013**, *2*, 856–862. [CrossRef]
34. Kim, D.Y.; Kim, H.S.; Kong, D.S.; Choi, M.; Kim, H.B.; Lee, J.-H.; Murillo, G.; Lee, M.; Kim, S.S.; Jung, J.H. Floating buoy-based triboelectric nanogenerator for an effective vibrational energy harvesting from irregular and random water waves in wild sea. *Nano Energy* **2018**, *45*, 247–254. [CrossRef]
35. Pu, X.; Liu, M.; Chen, X.; Sun, J.; Du, C.; Zhang, Y.; Zhai, J.; Hu, W.; Wang, Z.L. Ultrastretchable, transparent triboelectric nanogenerator as electronic skin for biomechanical energy harvesting and tactile sensing. *Sci. Adv.* **2017**, *3*, e1700015. [CrossRef] [PubMed]
36. Fan, F.-R.; Tian, Z.-Q.; Lin Wang, Z. Flexible triboelectric generator. *Nano Energy* **2012**, *1*, 328–334. [CrossRef]
37. Wang, Z.; Lin, L.; Chen, J.; Niu, S.; Zi, Y. *Triboelectric Nanogenerators*; Green Energy and Technology; Springer International Publishing: Berlin/Heidelberg, Germany, 2016; ISBN 978-3-319-40038-9.
38. Henniker, J. Triboelectricity in Polymers. *Nature* **1962**, *196*, 474. [CrossRef]
39. Chen, J.; Yang, J.; Li, Z.; Fan, X.; Zi, Y.; Jing, Q.; Guo, H.; Wen, Z.; Pradel, K.C.; Niu, S.; et al. Networks of Triboelectric Nanogenerators for Harvesting Water Wave Energy: A Potential Approach toward Blue Energy. *ACS Nano* **2015**, *9*, 3324–3331. [CrossRef]
40. Jiang, T.; Zhang, L.M.; Chen, X.; Han, C.B.; Tang, W.; Zhang, C.; Xu, L.; Wang, Z.L. Structural Optimization of Triboelectric Nanogenerator for Harvesting Water Wave Energy. *ACS Nano* **2015**, *9*, 12562–12572. [CrossRef]
41. Mariello, M.; Guido, F.; Mastronardi, V.M.; Todaro, M.T.; Desmaële, D.; De Vittorio, M. Nanogenerators for harvesting mechanical energy conveyed by liquids. *Nano Energy* **2019**, *57*, 141–156. [CrossRef]
42. Zi, Y.; Niu, S.; Wang, J.; Wen, Z.; Tang, W.; Wang, Z.L. Standards and figure-of-merits for quantifying the performance of triboelectric nanogenerators. *Nat. Commun.* **2015**, *6*, 8376. [CrossRef] [PubMed]
43. Niu, S.; Liu, Y.; Wang, S.; Lin, L.; Zhou, Y.S.; Hu, Y.; Wang, Z.L. Theoretical Investigation and Structural Optimization of Single-Electrode Triboelectric Nanogenerators. *Adv. Funct. Mater.* **2014**, *24*, 3332–3340. [CrossRef]
44. Phan, H.; Shin, D.-M.; Heon Jeon, S.; Young Kang, T.; Han, P.; Han Kim, G.; Kook Kim, H.; Kim, K.; Hwang, Y.-H.; Won Hong, S. Aerodynamic and aeroelastic flutters driven triboelectric nanogenerators for harvesting broadband airflow energy. *Nano Energy* **2017**, *33*, 476–484. [CrossRef]

45. Zhang, L.; Zhang, B.; Chen, J.; Jin, L.; Deng, W.; Tang, J.; Zhang, H.; Pan, H.; Zhu, M.; Yang, W.; et al. Lawn Structured Triboelectric Nanogenerators for Scavenging Sweeping Wind Energy on Rooftops. *Adv. Mater.* **2016**, *28*, 1650–1656. [CrossRef]
46. Chen, X.; He, J.; Song, L.; Zhang, Z.; Tian, Z.; Wen, T.; Zhai, C.; Chen, Y.; Cho, J.; Chou, X.; et al. Flexible one-structure arched triboelectric nanogenerator based on common electrode for high efficiency energy harvesting and self-powered motion sensing. *AIP Adv.* **2018**, *8*, 45022. [CrossRef]
47. Pu, X.; Guo, H.; Chen, J.; Wang, X.; Xi, Y.; Hu, C.; Wang, Z. Eye motion triggered self-powered mechnosensational communication system using triboelectric nanogenerator. *Sci. Adv.* **2017**, *3*, e1700694. [CrossRef]
48. Chen, B.D.; Tang, W.; He, C.; Deng, C.R.; Yang, L.J.; Zhu, L.P.; Chen, J.; Shao, J.J.; Liu, L.; Wang, Z.L. Water wave energy harvesting and self-powered liquid-surface fluctuation sensing based on bionic-jellyfish triboelectric nanogenerator. *Mater. Today* **2018**, *21*, 88–97. [CrossRef]
49. Quan, Z.; Han, C.B.; Jiang, T.; Wang, Z.L. Robust Thin Films-Based Triboelectric Nanogenerator Arrays for Harvesting Bidirectional Wind Energy. *Adv. Energy Mater.* **2016**, *6*, 1501799. [CrossRef]
50. Lin, Z.-H.; Cheng, G.; Li, X.; Yang, P.-K.; Wen, X.; Lin Wang, Z. A multi-layered interdigitative-electrodes-based triboelectric nanogenerator for harvesting hydropower. *Nano Energy* **2015**, *15*, 256–265. [CrossRef]
51. Wen, R.; Guo, J.; Yu, A.; Zhang, K.; Kou, J.; Zhu, Y.; Zhang, Y.; Li, B.-W.; Zhai, J. Remarkably enhanced triboelectric nanogenerator based on flexible and transparent monolayer titania nanocomposite. *Nano Energy* **2018**, *50*, 140–147. [CrossRef]
52. Arcot Narasimulu, A.; Zhao, P.; Soin, N.; Kovur, P.; Ding, P.; Chen, J.; Dong, S.; Chen, L.; Zhou, E.; Montemagno, C.; et al. Significant triboelectric enhancement using interfacial piezoelectric ZnO nanosheet layer. *Nano Energy* **2017**, *40*, 471–480. [CrossRef]
53. Kil Yun, B.; Soo Kim, H.; Joon Ko, Y.; Murillo, G.; Hoon Jung, J. Interdigital electrode based triboelectric nanogenerator for effective energy harvesting from water. *Nano Energy* **2017**, *36*, 233–240. [CrossRef]
54. Ravichandran, A.N.; Ramuz, M.; Blayac, S. Increasing surface charge density by effective charge accumulation layer inclusion for high-performance triboelectric nanogenerators. *MRS Commun.* **2019**, *9*, 682–689. [CrossRef]
55. Kim, H.-J.; Yim, E.-C.; Kim, J.-H.; Kim, S.-J.; Park, J.-Y.; Oh, I.-K. Bacterial Nano-Cellulose Triboelectric Nanogenerator. *Nano Energy* **2017**, *33*, 130–137. [CrossRef]
56. Zhang, M.; Xia, L.; Dang, S.; Shi, L.; Cao, A.; Deng, Q.; Du, C. A flexible single-electrode-based triboelectric nanogenerator based on double-sided nanostructures. *AIP Adv.* **2019**, *9*, 75221. [CrossRef]
57. Nguyen, V.; Kelly, S.; Yang, R. Piezoelectric peptide-based nanogenerator enhanced by single-electrode triboelectric nanogenerator. *APL Mater.* **2017**, *5*, 74108. [CrossRef]
58. Wang, M.; Zhang, N.; Tang, Y.; Zhang, H.; Ning, C.; Tian, L.; Li, W.; Zhang, J.; Mao, Y.; Liang, E. Single-electrode triboelectric nanogenerators based on sponge-like porous PTFE thin films for mechanical energy harvesting and self-powered electronics. *J. Mater. Chem. A* **2017**, *5*, 12252–12257. [CrossRef]
59. Gorham, W.F. A New, General Synthetic Method for the Preparation of Linear Poly-p-xylylenes. *J. Polym. Sci.* **1966**, *4*, 3027–3039. [CrossRef]
60. Lo, H.; Tai, Y.-C. Parylene-based electret power generators. *J. Micromechanics Microeng.* **2008**, *18*, 104006. [CrossRef]
61. Genter, S.; Langhof, T.; Paul, O. Electret-based Out-Of-Plane Micro Energy Harvester with Parylene-C Serving as the Electret and Spring Material. *Proc. Eng.* **2015**, *120*, 341–344. [CrossRef]
62. Jeong, G.S.; Baek, D.-H.; Jung, H.C.; Song, J.H.; Moon, J.H.; Hong, S.W.; Kim, I.Y.; Lee, S.-H. Solderable and electroplatable flexible electronic circuit on a porous stretchable elastomer. *Nat. Commun.* **2012**, *3*, 977. [CrossRef] [PubMed]
63. Priya, S. Modeling of electric energy harvesting using piezoelectric windmill. *Appl. Phys. Lett.* **2005**, *87*, 184101. [CrossRef]
64. Zou, H.; Zhang, Y.; Guo, L.; Wang, P.; He, X.; Dai, G.; Zheng, H.; Chen, C.; Wang, A.; Xu, C.; et al. Quantifying the triboelectric series. *Nat. Commun.* **2019**, *10*, 1–9. [CrossRef] [PubMed]
65. Xiong, J.; Cui, P.; Chen, X.; Wang, J.; Parida, K.; Lin, M.-F.; Lee, P.S. Skin-touch-actuated textile-based triboelectric nanogenerator with black phosphorus for durable biomechanical energy harvesting. *Nat. Commun.* **2018**, *9*, 1–9. [CrossRef] [PubMed]

66. Chen, T.-N.; Wuu, D.-S.; Wu, C.-C.; Chiang, C.-C.; Chen, Y.-P.; Horng, R.-H. Improvements of Permeation Barrier Coatings Using Encapsulated Parylene Interlayers for Flexible Electronic Applications. *Plasma Process. Polym.* **2007**, *4*, 180–185. [CrossRef]
67. Li, W.; Rodger, D.; Menon, P.; Tai, Y.C. Corrosion Behavior of Parylene-Metal-Parylene Thin Films in Saline. *ECS Trans.* **2008**, *11*, 1. [CrossRef]
68. Mariello, M.; Guido, F.; Mastronardi, V.M.; Giannuzzi, R.; Algieri, L.; Qualteri, A.; Maffezzoli, A.; De Vittorio, M. Reliability of Protective Coatings for Flexible Piezoelectric Transducers in Aqueous Environments. *Micromachines* **2019**, *10*, 739. [CrossRef]
69. Mariello, M.; Guido, F.; Mastronardi, V.M.; De Donato, F.; Salbini, M.; Brunetti, V.; Qualtieri, A.; Rizzi, F.; De Vittorio, M. Captive-air-bubble aerophobicity measurements of antibiofouling coatings for underwater MEMS devices. *Nanomater. Nanotechnol.* **2019**, *9*. [CrossRef]

energies

MDPI

Article

A Minimal Volume Hermetic Packaging Design for High-Energy-Density Micro-Energy Systems †

Xiujun Yue, Jessica Grzyb, Akaash Padmanabha and James H. Pikul *

Department of Mechanical Engineering and Applied Mechanics, University of Pennsylvania, Philadelphia, PA 19104, USA; xiujun@seas.upenn.edu (X.Y.); jagrzyb@seas.upenn.edu (J.G.); apadman@seas.upenn.edu (A.P.)

* Correspondence: pikul@seas.upenn.edu

† This paper is an extended version of our paper published in Proceedings of PowerMEMS 2019, Kraków, Poland, 2–6 December 2019; pp. 1–4, doi:10.1109/PowerMEMS49317.2019.30773702286.

Received: 27 April 2020; Accepted: 14 May 2020; Published: 15 May 2020

Abstract: Hermetic packaging is critical to the function of many microscale energy storage and harvesting devices. State-of-the-art hermetic packaging strategies for energy technologies, however, are designed for macroscale devices and dramatically decrease the fraction of active materials when applied to micro-energy systems. We demonstrated a minimal volume hermetic packaging strategy for micro-energy systems that increased the volume of active energy storage materials by 2× and 5× compared to the best lab scale microbatteries and commercial pouch cells. The minimal volume design used metal current collectors as a multifunctional hermetic shell and laser-machined hot melt tape to provide a thin, robust hermetic seal between the current collectors with a stronger adhesion to metals than most commercial adhesives. We developed the packaging using commercially available equipment and materials, and demonstrated a strategy that could be applied to many kinds of micro-energy systems with custom shape configurations. This minimal, versatile packaging has the potential to improve the energy density of current micro-energy systems for applications ranging from biomedical devices to micro-robots.

Keywords: hermetic packaging; IoT; MEMS; microbattery; battery; harvester

1. Introduction

Hermetic packaging technologies that prevent internal components from reacting with oxygen or moisture in the air are critical for numerous microscale technologies, including sensors, batteries, super-capacitors, energy harvesters and other energy systems [1–3]. Creating suitable packaging strategies for these microscale technologies is of growing importance as the markets for these devices continue to increase. The microbattery market, for example, is predicted to grow nearly 5× between 2019 and 2025 as a result of new Internet of Things (IoT) and medical devices [4], but current hermetic packaging technologies limit microbattery energy densities to a fraction of macroscale batteries.

One reason for the divergent energy densities of micro- and macroscale batteries is that widely used macroscale hermetic packaging technologies cannot be directly applied to microbatteries as the packaging dominates the volume and mass of the internal components [5–7]. Battery pouch cells, for example, have a minimum hot sealing width of 3 mm, which accounts for 12% of the total cell area in a 10 × 10 cm^2 macroscale battery, and allows 88% of the cell area to be available for energy storing materials. When the same pouch cell packaging is applied to a 1 × 1 cm^2 microscale cell, the seal uses 84% of the cell area and leaves only 16% of the area for energy storing materials. If the energy storing materials had a combined energy density of 1000 Wh/L, the macroscale cell would have 880 Wh/L energy density, while the microscale cell would have only 160 Wh/L.

Due to extreme difficulties in realizing low-volume fraction microscale packaging, the field lacks available literature, and the majority of micro-energy systems studied have been demonstrated without packaging [7–14]. A microbattery by W. Lai et al. is one of the very few examples with packaging, but the bulky sealing left only 45% of the cell volume available for electrodes, which limited the full cell energy density to about 400 Wh/L [15]. Additional prior work reduced the packaging fraction by using current collectors as part of the packaging and demonstrated the advantage of this approach [15,16]. In addition to requiring a low volume fraction, practical microscale hermetic packaging technologies need to be mechanically strong, chemically inert and easily assembled at scale.

Here, we presented a microscale hermetic packaging strategy that decreased the packaging volume fraction using two innovations: (1) the metal current collectors served as part of the hermetic packaging, and (2) ultra-thin strips of laser-machined hot melt tape hermetically sealed the gap between the current collectors. Figure 1 shows the packaging structure using a li-ion battery as an example. The volume consumed by packaging in this battery was less than 20%. This packaging volume fraction provided over a 2× improvement compared to the best lab scale microbatteries, and over a 3× improvement compared to the commercially available li-ion batteries, where packaging volume consumes as much as 70% [17]. Furthermore, with 84% of the 1 × 1 cm^2 cell area available for active material, this strategy achieved a 5× improvement in the amount of volume available for energy storage materials compared to macroscale pouch cell sealing.

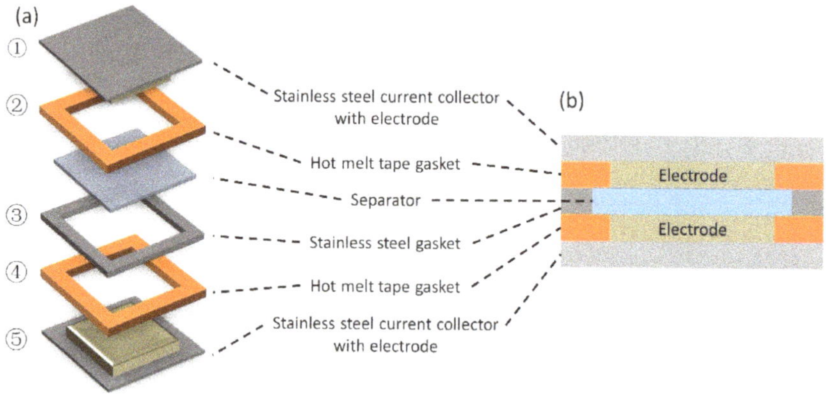

Figure 1. Schematic of the hermetic microscale packaging (**a**) component breakdown and (**b**) cross section view.

In addition to reducing the packaging volume fraction, the packaging design maintains impressive mechanical strength and stability. T-peel tests showed that the packaging structure exhibited an ultra-high adhesive strength of 43.10 N/cm, which was twice the strength of commercial pouch cells and over two times the strength of commonly used commercial metal adhesives. Additionally, the results from cyclic voltammetry and a high voltage tests showed that the packaging structure exhibited impressive electrochemical stability, with no reaction detected at up to 5 V. The presented packaging strategy offered both a significant volumetric energy density advantage and a robust, hermetic seal for micro-energy systems.

2. Materials and Methods

For the micro-device packaging, we fabricated 1 × 1 cm^2 metal current collectors using laser micromachining equipment (IPG Microsystems IX-200-F, laser at 532 nm, 10 kHz; the power and pulse number varied based on different materials and thicknesses). We machined stainless steel, copper and aluminum with thicknesses between 10 μm to 50 μm. The resulting 1 × 1 cm^2 metal squares represented current collectors that could be used in any microsystem, for instance as current collectors in a li-ion battery or supercapacitor on which electrode materials could be deposited.

Hot melt tape gaskets were used to create a strong hermetic seal between the metal current collectors. We made tape gaskets from 100 μm thick enhanced sulfurized polymer resin tri-layer hot melt tape from the MTI Corporation (PLIB-HMA30). Polymer resin hot melt tape was designed to seal the pouch cell cases with metal terminal tabs. Being designed for battery applications, the material has strong adhesion to metal, good electrolyte stability and low air and water permeability. An excimer laser (IPG Microsystems IX-255) with a 193 nm ultraviolet light source machined the optically transparent hot melt tape into 1 × 1 cm^2 windows with 400 μm wide boarders (Figure 2a). The 10 μm laser resolution produced tape gaskets with only a ±5% error in edge length. By using solid gaskets and high-precision micromachining equipment, we enabled a low-volume fraction seal with significantly better dimensional precision than liquid- or gel-based adhesives.

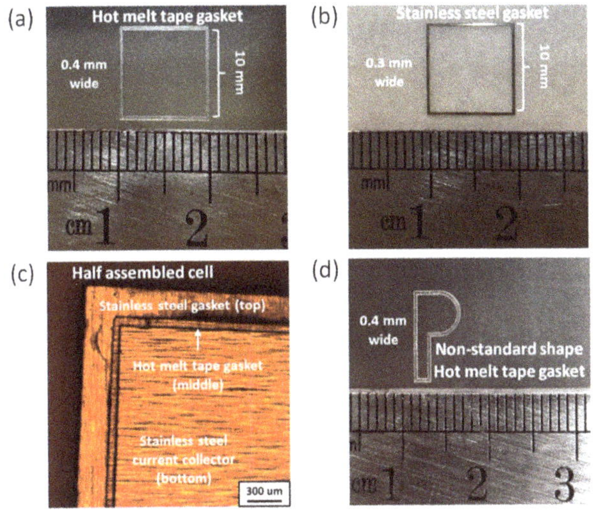

Figure 2. (**a**) Optical image of a precision-machined hot melt tape gasket. (**b**) Optical image of a precision-machined stainless steel gasket. (**c**) Optical microscope image of a half assembled cell with a hot melt tape gasket adhered to a stainless steel current collector (bottom) and a stainless steel gasket (top). (**d**) An example of micro-packaging with non-standard shape.

A micro-machined hot melt tape gasket was bonded between two metal current collectors to create a 1 × 1 cm^2 package for micro-energy systems. The hermetic seal was completed by hot pressing each side of the 1 × 1 cm^2 package using a commercially available impulse heat sealer (FS-300, 430W, 12-inch-long, 3 mm wide filament). To achieve a strong adhesion, the tape should reach 135 °C at 0.3 MPa for 2–3 s, depending on the current collector. The heat setting was adjusted on the impulse sealer and the pressure was adjusted by the spring in the sealer. To adhere the hot melt tape onto a 50 μm thick stainless steel foil, we set the heating filament to 180 °C for 3 s which ensured that enough heat penetrated the foil to bring the adhesive to 135 °C. The heat adhered the inner hot melt tape gasket to the surrounding metal components and once cooled, created an ultra-strong solid adhesive bond.

A unique advantage to this packaging design was that the thickness could be easily adjusted for different micro-devices. To achieve higher thicknesses, we stacked multiple tape gaskets with an intermediate metal gasket between each consecutive tape layer. We obtained up to 300 μm thick gaskets, though greater thicknesses (>mm) were achievable. For the stacked layers, we fabricated metal gaskets using laser-micro-machining equipment, however, the metal gaskets had 300 μm wide windows (100 μm less than the tape gasket) so the metal would avoid contact with an inner electrode material in a battery or micro-device (Figure 2c). Additionally, we modified the hot sealing sequence to ensure good adhesion between all the layers. We started by stacking a current collector (① in

Figure 1), a tape gasket (②) and a stainless steel gasket (③) and hot pressing the three components together (Figure 2c). We then hot pressed on one additional tape gasket (④) with a stainless steel gasket (③′, same as ③), following a metal–polymer–metal sequence, and repeated these incremental additions until the desired thickness was achieved. The advantage to adding components in sequential steps, and never heating more than one new metal/tape/metal combination was to create easier an alignment of the packaging components and optimize the heat transfer to the tape from the impulse sealer's one-sided heat source. After adding the intermediate stainless steel and tape layers, we completed the final step by hot pressing the remaining current collector (⑤) to the multilayer structure.

The mechanical and electrochemical properties of the micro-packaging strategy were evaluated and compared to other packaging designs. T-peel experiments tested the mechanical sealing strength between the hot melt tape and the metal current collectors. Different adhesives were used to bond two identical stainless steel strips (10 cm × 1 cm × 50 μm) at one end. Two free ends were mounted onto a mechanical tester (Instron 5564, speed at 5 mm/min), which tested the load required to peel the strips apart (Figure 3). One sample was tested for each data point. Cyclic voltammetry scans between 0–5 V and a potentiostatic test at 5 V were performed on a dummy cell, which is a fully packed cell without electrode materials and electrolyte, to test the packaging's electrochemical stability. The sample set up was the same as the T-peel test as shown in Figure 3a. The bonded part was 1 × 1 cm^2 and two free ends of stainless steel strips were connected to a potentiostat (BioLogic VMP-300).

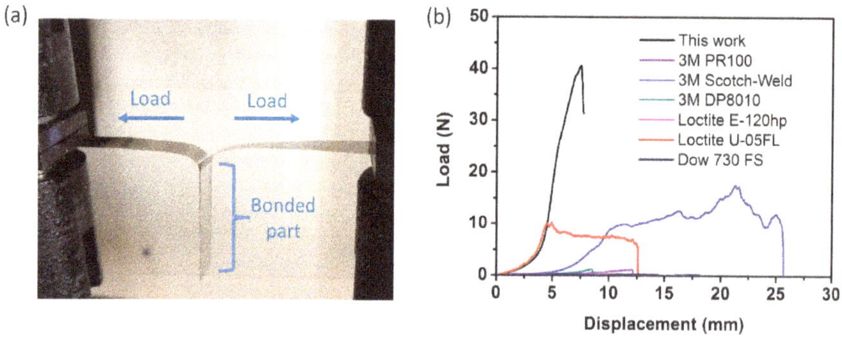

Figure 3. Peel strength of adhesives. (**a**) T-peel test setup and (**b**) the load-displacement curves of the T-peel tests for the different adhesives on steel.

The permeability of the micro-packaging strategy was studied via the stability of lithium in a dummy cell. One piece of 0.5 × 0.5 cm^2 lithium foil was sealed in a 1 × 1 cm^2 dummy cell made with 50 μm thick stainless steel foils and 400 μm wide hot melt tape gasket inside an argon-filled glovebox (H_2O, O_2 < 1 ppm). After storing the cell at ambient condition for two months, we opened the cell and checked the lithium surface for changes.

3. Results and Discussion

The durability of stainless steel, copper and aluminum current collectors during the packaging assembly were tested. The metal current collectors provided excellent hermetic seals. Stainless steel current collectors had the highest durability and did not fracture during assembly even at a low 10 μm thickness. In comparison, thin copper and aluminum foils were more susceptible to coiling and tearing. Without electrode materials to mechanically reinforce the current collectors, copper and aluminum required greater than 50 μm thicknesses to avoid failure. Lower thicknesses were possible when combined with electrode materials. With the aim of minimizing the material volume, we determined 10 μm stainless steel foils were optimal current collectors for our 1 × 1 cm^2 packaging.

The T-peel tests showed that the adhesion between the tape gaskets and current collectors exhibited an ultra-high strength of 43.10 N/cm. Table 1 compares the strength of this packing structure with other materials commonly used in electronic device packaging. As illustrated, this work presented a packaging strategy with twice the peel strength of commercial pouch cell cases and significantly higher than the popular industrial adhesives.

Table 1. Peel strength of the different adhesives on stainless steel.

	This Work	3M PR100	3M Scotch-Weld	3M DP8010	Loctite E-120hp	Loctite U-05FL	Dow 730 FS
Peel strength (N/cm)	43.1	1.3	17.7	1.3	0.24	11.8	0.25

We assessed the electrochemical stability of the packaging structure through cyclic voltammetry and constant high voltage tests (Figure 4a,b). We used a dummy cell, in which there was no active material on the metal current collectors, to test the electrochemical response of the packaging components only. The experiment tested the stability between 0 to 5 V. The detected currents were in the range of equipment background and noise, which also proved the insulating capability of the sealing structure. The results from the CV and high-voltage test showed that the packaging structure exhibited impressive electrochemical stability, with no reaction detected even at 5 V.

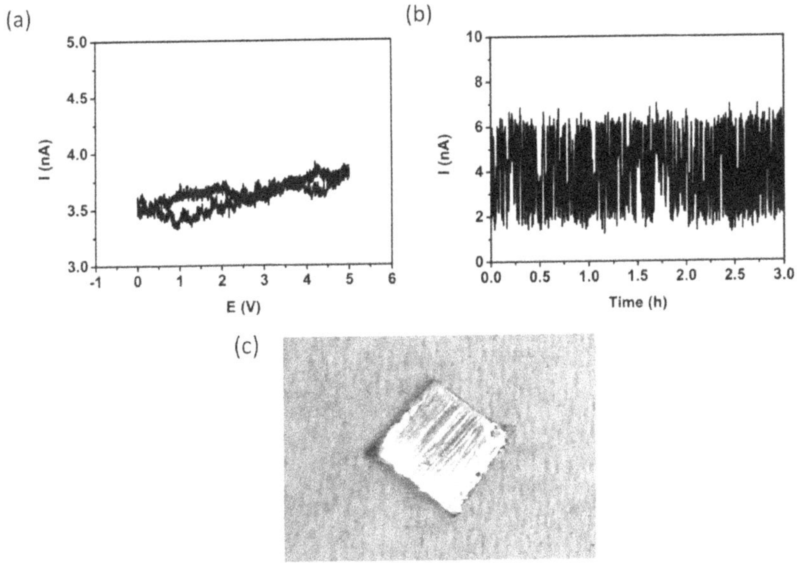

Figure 4. Stability of the micro-packaging. (**a**) Cyclic voltammetry. Scan rate: 20 mV/s (**b**) Potentiostatic test at 5 V. The voltage was applied over the stainless steel/tape/stainless sealing in a fully assembled dummy cell. (**c**) The lithium foil after storing inside a micro-package at ambient condition for two months.

The lithium stability test also demonstrated the excellent hermetic property of the packaging structure. Figure 4c shows the lithium sealed inside the dummy cell and stored in ambient condition for two months. The lithium kept its original shinny metallic surface which indicated that no detectable water or oxygen leaked through the 400 µm wide hot melt tape gasket.

We developed a low volume fraction micro-packaging strategy designed to increase the volumetric and gravimetric energy density of micro-energy storage technologies. This approach improved upon standard MEMS packaging solutions where the total area, but not necessarily weight and volume, are important [15]. In addition, our approach can be applied with moderate temperatures and is compatible with organic and corrosive solvents. Many MEMS-packaging solutions, in contrast, use glass and silicon bonding which require very high temperatures that adversely affect the critical microbattery components, like the separator, and can cause phase changes in the active electrodes [18]. Many microfluidic device-packaging strategies have also been designed to contain fluids, but common materials like polydimethylsiloxane (PDMS) and epoxies are not compatible with the organic solvent and corrosive salts in energy technologies [19]. Our packaging strategy overcomes the material and processing limitations of prior packaging solutions [20].

4. Conclusions

We demonstrated a micro-packaging strategy that is mechanically and chemically robust using commercially available equipment and materials. The minimal volume packaging design incorporated metal current collectors as part of the hermetic packaging and achieved an ultra-strong seal between the current collectors using thin hot melt tape gaskets. This strategy provided a 5× increase in the area available for active materials compared to pouch cells and a 2× reduction in packaging volume compared to the best lab scale microbatteries. The design is versatile in both shape and thickness, and has potential to be configured for a variety of micro-energy storage systems. The advancements provided by this micro-packaging strategy can be used to improve the energy density of current micro-energy storage technologies for a wide range of applications.

Author Contributions: Conceptualization, X.Y. and J.H.P.; methodology, X.Y.; investigation, X.Y., J.G. and A.P.; visualization, X.Y. and A.P.; writing—original draft preparation, J.G.; writing—review and editing, X.Y and J.H.P.; supervision, J.H.P.; funding acquisition, J.H.P. All authors have read and agreed to the published version of the manuscript.

Funding: This research was funded by the DARPA SHRIMP contract # HR0011-19-C-0056, and the work was carried out in part at the Singh Center for Nanotechnology, which is supported by the NSF National Nanotechnology Coordinated Infrastructure Program under grant NNCI-1542153.

Conflicts of Interest: The authors declare no conflict of interest.

References

1. Lee, Y.C.; Kong, M.; Zhang, Y. Microelectromechanical Systems and Packaging. In *Materials for Advanced Packaging*, 2nd ed.; Lu, D., Wong, C., Eds.; Springer: Cham, Switzerland, 2017.
2. Qian, Y.; Lee, C. System Packaging and Assembly in IoT Nodes. In *Enabling the Internet of Things*; Springer Science and Business Media LLC: Berlin/Heidelberg, Germany, 2017; pp. 441–482.
3. Pikul, J.H.; Ning, H. Powering the Internet of Things. *Joule* **2018**, *2*, 1036–1038. [CrossRef]
4. Micro Battery Market. Available online: www.marketsandmarkets.com/Market-Reports/micro-battery-market-25755945.html (accessed on 27 March 2020).
5. Dillon, S.J.; Sun, K. Microstructural design considerations for Li-ion battery systems. *Curr. Opin. Solid State Mater. Sci.* **2012**, *16*, 153–162. [CrossRef]
6. Liu, L.; Weng, Q.; Lu, X.; Sun, X.; Zhang, L.; Schmidt, O.G. Advances on Microsized On-Chip Lithium-Ion Batteries. *Small* **2017**, *13*, 1701847. [CrossRef]
7. Hur, J.I.; Smith, L.C.; Dunn, B. High Areal Energy Density 3D Lithium-Ion Microbatteries. *Joule* **2018**, *2*, 1187–1201. [CrossRef]
8. Pikul, J.H.; Zhangb, H.; Cho, J.; Braun, P.V.; King, W.P. High-power lithium ion microbatteries from interdigitated three-dimensional bicontinuous nanoporous electrodes. *Nat. Commun.* **2013**, *4*, 1732. [CrossRef] [PubMed]
9. Cheah, S.K.; Perre, E.; Rooth, M.; Fondell, M.; Hårsta, A.; Nyholm, L.; Boman, M.; Gustafsson, T.; Lu, J.; Simon, P.; et al. Self-Supported Three-Dimensional Nanoelectrodes for Microbattery Applications. *Nano Lett.* **2009**, *9*, 3230–3233. [CrossRef] [PubMed]
10. Golodnitsky, D.; Yufit, V.; Nathan, M.; Shechtman, I.; Ripenbein, T.; Strauss, E.; Menkin, S.; Peled, E. Advanced materials for the 3D microbattery. *J. Power Sources* **2006**, *153*, 281–287. [CrossRef]

11. Ren, J.; Li, L.; Chen, C.; Chen, X.; Cai, Z.; Qiu, L.; Wang, Y.; Zhu, X.; Peng, H. Batteries: Twisting Carbon Nanotube Fibers for Both Wire-Shaped Micro-Supercapacitor and Micro-Battery (Adv. Mater. 8/2013). *Adv. Mater.* **2013**, *25*, 1224. [CrossRef]
12. Wang, K.; Zou, W.; Quan, B.; Yu, A.; Wu, H.; Jiang, P.; Wei, Z. An All-Solid-State Flexible Micro-supercapacitor on a Chip. *Adv. Energy Mater.* **2011**, *1*, 1068–1072. [CrossRef]
13. Beidaghi, M.; Wang, C. Micro-supercapacitors based on three dimensional interdigital polypyrrole/C-MEMS electrodes. *Electrochim. Acta* **2011**, *56*, 9508–9514. [CrossRef]
14. Liu, W.; Yan, X.; Chen, J.; Feng, Y.; Xue, Q. Novel and high-performance asymmetric micro-supercapacitors based on graphenequantum dots and polyaniline nanofibers. *Nanoscale* **2013**, *5*, 6053–6062. [CrossRef] [PubMed]
15. Lai, W.; Erdonmez, C.K.; Marinis, T.F.; Bjune, C.K.; Dudney, N.J.; Xu, F.; Wartena, R.; Chiang, Y.-M. Ultrahigh-Energy-Density Microbatteries Enabled by New Electrode Architecture and Micropackaging Design. *Adv. Mater.* **2010**, *22*, E139–E144. [CrossRef] [PubMed]
16. Kim, H.; Auyeung, R.C.Y.; Pique, A. Laser-printed thick-film electrodes for solid-state rechargeable Li-ion microbatteries. *J. Power Sources* **2007**, *165*, 413–419. [CrossRef]
17. Bjune, C.K.; Marinis, T.F.; Chiang, Y.-M. Packaging Design and Assembly for Ultrahigh Energy Density Microbatteries. *J. Microelectron. Electron. Packag.* **2015**, *12*, 129–138. [CrossRef]
18. Yang, H.-A.; Wu, M.; Fang, W. Localized induction heating solder bonding for wafer level MEMS packaging. *J. Micromech. Microeng.* **2004**, *15*, 394–399. [CrossRef]
19. Nge, P.N.; Rogers, C.I.; Woolley, A.T. Advances in Microfluidic Materials, Functions, Integration, and Applications. *Chem. Rev.* **2013**, *113*, 2550–2583. [CrossRef] [PubMed]
20. Yue, X.; Grzyb, J.; Padmanacha, A.; Pikul, J. A minimal volume hermetic packaging design for high energy density micro energy systems. In Proceedings of the 2019 19th International Conference on Micro and Nanotechnology for Power Generation and Energy Conversion Applications (PowerMEMS), Kraków, Poland, 2–6 December 2019; Institute of Electrical and Electronics Engineers (IEEE): Piscataway, NJ, USA; pp. 1–4.

energies

MDPI

Article

Inkjet 3D Printed MEMS Vibrational Electromagnetic Energy Harvester

Bartosz Kawa [1], Krzysztof Śliwa [1], Vincent Ch. Lee [2], Qiongfeng Shi [2] and Rafał Walczak [1,*]

1 Department of Microsystems, Wroclaw University of Science and Technology, 50370 Wroclaw, Poland; bartosz.kawa@pwr.edu.pl (B.K.); krzysztof.sliwa@pwr.edu.pl (K.Ś.)

2 Department of Electrical and Computer Engineering, National University of Singapore, Singapore 117608, Singapore; elelc@nus.edu.sg (V.C.L.); eleshiq@nus.edu.sg (Q.S.)

* Correspondence: rafal.walczak@pwr.edu.pl; Tel.: +34-48-320-4973

Received: 6 May 2020; Accepted: 1 June 2020; Published: 1 June 2020

Abstract: Three-dimensional (3D) printing is a powerful tool that enables the printing of almost unlimited geometry in a few hours, from a virtual design to a real structure. In this paper, we present a micro-electromechanical energy harvester that utilized a 3D printed micromechanical structure combined with a miniature permanent magnet and a microelectronic coil towards a hybrid electromagnetic vibrational hybrid energy harvester. Various micromechanical structure geometries were designed, printed, and tested. The characteristic dimensions of the springs were from 200 µm to 400 µm and the total volume of the devices was below 1 cm^3. The resonant frequencies (95–340 Hz range), as well as bandwidths (6–23 Hz range), for the developed prototypes were determined. The maximal generated output power was almost 24 µW with a power density up to almost 600 $\mu W/cm^3$.

Keywords: 3D printing; MEMS; energy harvester

1. Introduction

Micro-electromechanical systems (MEMS) combine mechanical microstructures with microelectronic circuits to create a very small functional system that senses, actuates, or harvests energy. The technology of the micromechanical structures in MEMSs involves applying many well-known techniques to the specific materials. Silicon and glass are micromachined mainly by wet or dry etching [1–3], polymer microstructures are formed by injection molding, hot embossing, or soft lithography [4–7], and low-temperature co-fired ceramic substrates are cut and co-fired [8]. Regardless of the applied material and technology, the fabrication of MEMSs is a multistep process involving many technological steps (photolithography, etching, deposition, bonding, assembling, etc.) that require specialized facilities (i.e., apparatus and clean rooms), trained staff, and often knowledge on the properties of the applied materials and limits of the used techniques that are collected over years of experience. All these issues mean that although single MEMSs are usually low-cost devices, a further decrease in the cost-per-chip is difficult to achieve. This is important because some of the forecasts from established companies and institutions (including Bosch, HP, Cisco, and Intel) clearly indicate that in the next decade, the number of MEMS-based devices connected to the internet (Internet of Things, IoT) is going to reach the level of trillions [9]. From an economical point of view, this requires a decrease in the cost of networked MEMSs by at least one or two orders of magnitude. A low price and short development time would result in a higher availability of networked MEMSs for the IoT approach. One promising technology that may decrease the fabrication costs of MEMSs is additive manufacturing, commonly known as 3D printing. Various 3D printing techniques (for example, inkjet printing, stereolithography, or fused deposition modeling) were already successfully applied to develop some MEMS devices. The exponential growth of scientific papers on 3D-printed microfluidic chips is observed during the last decade [10–14]. Additive manufacturing enables the fast and cost-effective development of truly three-dimensional microfluidic structures. Additionally, a boom in 3D printed sensors is widely reported [15]. A discussion on the applicability of inkjet 3D printing in MEMS technology was recently presented by Walczak [16]. In addition, some examples of 3D printed microdevices, including pumps [17], valves [18], flow regulators [19], electrostatic actuators [20], fuze [21], or specialized packages for MEMS devices [22]. Thus, 3D printing seems to be an attractive alternative for standard microengineering techniques applied in MEMS technology.

The energy source for the power supply of MEMS-based sensing nodes in IoT systems is a crucial issue in decreasing the cost. A comprehensive review on MEMS vibrational energy harvesters was recently presented by Toshiyoshi et al. [23]. After years of miniature energy harvester development, more than a few mW of power can be gained for the power supply of low-power electronic circuits using photovoltaic cells or thermoelectric generators. Mechanical vibration is an attractive candidate as an energy source because is it available almost everywhere. Thus, electromagnetic [24], piezoelectric [25], triboelectric [26], or electrostatic [27] MEMS energy harvesters are reported in the literature. Those devices produce electric energy from a few μW to a few mW. They are made from traditional materials (silicon glass, polymers, metals) using microelectronic and microengineering techniques related to MEMS technology. Only a few of them are utilizing 3D printing technology as a fabrication tool.

The first reported applications of 3D printing were for the housing of the energy harvester. Due to the required low precision of the housing fabrication, a fused deposition modeling (FDM) technique was applied [28,29]. Ju et al. described an impact-based piezoelectric vibrational energy harvester that had FDM printed housing with $26 \times 10 \times 10$ mm^3 dimensions [28]. Maharjan et al. presented a 3D-printed hybridized electromagnetic–triboelectric nanogenerator where two half parts of the housing (circular tubes with a 1.45 cm inner diameter and an outer diameter of a few cm) were 3D printed [29]. The next step in the evolution of 3D-printed energy harvesters was the fabrication of the active mechanical elements of the energy harvesters. Seol et al. described all 3D-printed triboelectric energy harvesters [30]. A grating disk type triboelectric nanogenerator was fabricated by assembling a 3D-printed electrode layer, a triboelectric layer, and a case package. In this solution, three printing techniques—metal printing, FDM, and stereolithography (STL)—were successfully applied. The overall dimensions of the device were counted in cm. The achieved power density was 4.52 mW/cm^3. The same group of researchers also developed a triboelectric nanogenerator consisting of a 3D-printed case, a core oscillator, and a spring, all printed using the FDM technique. The harvester was 90 mm long, 30 mm wide, and 15 mm high. It produced a maximum voltage of 98.2 V and current of 13.7 μA [31]. Electromagnetic energy harvesters have also been developed with the use of 3D printing. Hadas et al. designed and tested an energy harvester with FDM-printed immovable elements of the device [32]. The dimensions of the harvester were $8 \times 6 \times 6$ cm^3 and the maximal generated power was equal to 26 mW (power density 90 μW/cm^3). Immovable elements printed using FDM were also successfully applied in the energy harvester developed by Rues et al. [33]. The harvester had the dimensions of $4.5 \times 5.5 \times 4$ cm^3, and 127 mW was measured as the maximum possible generated power (power density 1.28 mW/cm^3). Spherical magnetic generators for bio-motional energy harvesting were developed by Bowers et al. [34]. Two semispherical energy harvester structures with a total diameter of 1.91 cm were inkjet printed and glued together. The maximal generated output power was 1.44 mW (power density around 0.4 μW/cm^3). A nonlinear vibrational energy harvester was described by Constantinou et al. [35]. The dimensions of the printed harvester were around $6 \times 4 \times 2$ cm^3, but the vibrating double beam was 17.8 mm high, with a single beam thickness of 0.8 mm and a width of 4 mm. The maximal generated power density was 0.48 mW/cm^3. Only in a few of the reports listed above were 3D-printed movable mechanical components coupled with an electromagnetic circuit operating at the resonance of the oscillating mechanism. The dimensions of most of the printed harvesters were in cm, with a volume well above 1 cm^3. The FDM technique dominates as the technique for printing mechanical elements due to its simplicity and the low costs of the printing material and device.

In this paper, we demonstrate the use of inkjet 3D printing as a technique that enables the fabrication of micro-electromechanical electromagnetic energy harvesters with a volume below 1 cm^3. We consider simple micro-mechanic structures with a seismic mass suspended on springs with different geometrical dimensions. Numerical simulations of the resonance frequencies are followed by measurements. A critical issue is the measurement of the real mechanical properties of the investigated structures as input data for simulations. Finally, the electrical characteristics of the 3D-printed harvesters, as well as the generated power as a function of the vibration frequency and resistive load, are presented and discussed.

2. Materials and Methods

2.1. Design and Numerical Simulations

The harvesters were designed using Inventor software (Autodesk, San Rafa, CA, USA). The external dimensions of the device were the same for all considered internal designs (Figure 1). The internal structure consisted of a seismic mass with a place for small commercial permanent neodymium magnets (1.5 mm in diameter, 1 mm high, Webcraft, Germany). One or two magnets were positioned on the sides of the vibrating microstructure. The seismic mass was suspended on springs (various designs) with thicknesses of 200 µm or 400 µm. The internal structure volume, called the active volume [35], was taken into account in the calculation of the generated power density. At a distance of 2 mm above the magnets, a commercially available miniature surface-mounted coil (SMD 812 with 7.36 mH inductance) was positioned in a printed holder. The output voltage from the coil was connected to a measurement setup. The scheme of the harvester and a computer visualization of one of the designed harvesters with a magnet and coil is shown in Figure 1. Simulations of the natural resonance frequency of the vibrating structure with magnets were carried out using Inventor software (Autodesk, San Rafa, CA, USA). The simulations were conducted for an acrylonitrile butadiene styrene-like material with a Young's modulus equal to the printing material's data sheet (E = 1423 MPa) and test structures measured by us.

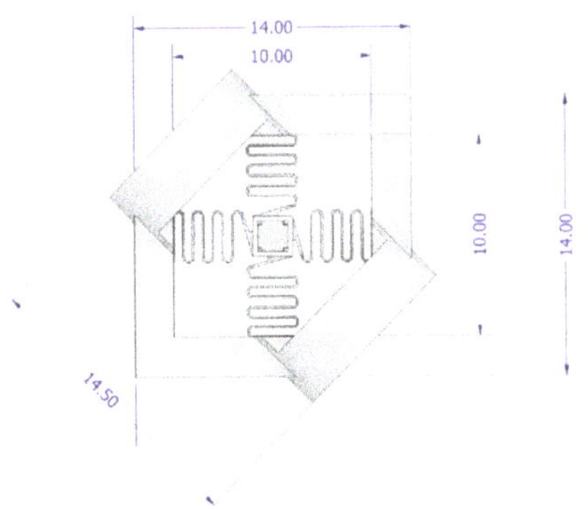

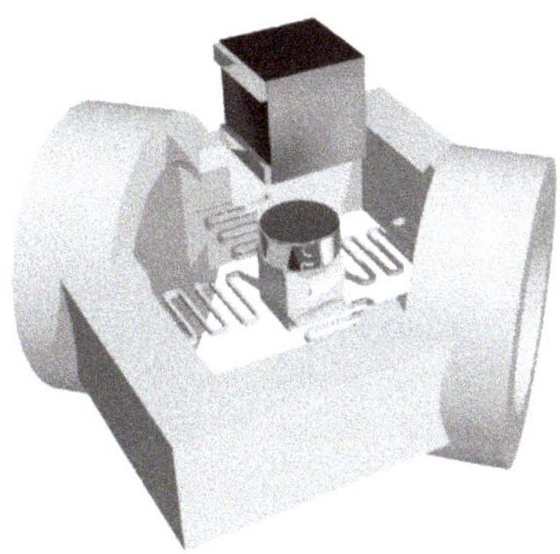

Figure 1. (**left**) Visualization of the 3D-printed energy harvester with main dimensions in mm and (**right**) with assembled magnet coupled with one miniature coil.

2.2. Inkjet 3D Printing

The devices were fabricated using inkjet 3D printing. Harvester structures were printed using a 3D Systems ProJet 3510 printer with ultra-high definition resolution (single layer thickness equal to 16 µm and planar X–Y resolution of 750 dpi). Two standard inkjet printing materials, Visijet M3 Crystal (as building material) and S300 (as support material), were utilized. The layer structure of the harvester (including the spring structure) was printed parallel to the printing platform. This is an important printing parameter due to the different mechanical properties of microstructures with various layers of internal structure [36,37]. Post-processing was carried out according to a procedure developed by us and described in [38]. Briefly, raw printed structures with support material were put in an oven (2.5 h at 60 °C) to melt away the support material. The structures were cleaned in mineral oil at 60 °C with ultrasonic agitation (1 h). Finally, the structures were rinsed with deionized water and dried in a stream of dry air.

2.3. Determination of Young's Modulus

It is well known from our previous works that printed microstructures with dimensions like those considered in this paper have a different Young's modulus than reported in the material's data sheet. Therefore, knowledge of the real material properties is crucial for simulation parameters. In order to measure the real Young's modulus of the microstructures, ten microbeams with a constant length (6 mm) and 200 μm or 400 μm width and thickness were printed. They were printed in the same process, with the same dimensions, and mechanically characterized. The mechanical characterization of the microbeams was carried out according to the procedure and apparatus applied by us earlier, as described in [37]. The procedure involved a Bondtester Dage 4000 Plus (Nordson Dage, Feldkirchen, Germany) for microbeam deflection versus applied bending force. The deflection of the microbeam was measured with ±1 μm accuracy. The maximal deflection was set to 1.25 mm and the maximal applied force was 70 mN (limited by the device settings). In order to investigate the potential aging of the loaded microstructure, the microbeams were deflected 100 times. The values of deflection for the first and last loading tests for 10% and 90% of the maximal applied force were determined and used to calculate Young's modulus according to the following equation (concentrated force at the free end) [37] (1):

$$E = (4 \times F \times l^3)/(f \times b \times h^3) \quad (1)$$

where F–applied force, l–length of the microbeam, f–deflection of the microbeam, b–width of the microbeam, and h–thickness of the microbeam.

2.4. Measurement Setup

The measurement setup consisted of a vibration controller Type 2718, a power amplifier Type 2718, a vibration exciter 4809 (10–20 kHz) with a set constant acceleration (1 g), a digital oscilloscope, a laser source (635 nm, 5 mW) with a power supplier, and a photodetector (OP101, Thorlabs) (Figure 2). The electrical output of the harvester was connected to a loading resistance in the range of 10 Ω to 2.4 kΩ to measure output power vs. loading resistance characteristics. Despite collecting the electrical signal from the coil, the optical detection systems (laser and photodetector) were used to determine the resonant frequency of the vibrating microstructure. A charge coupled device (CCD) camera with a video grabber and USB connection (2 megapixels, HDCE-X2, Poland) observed the harvester during operation to monitor the operation of the vibration structure and determine the displacement of the vibrating seismic mass. It captured images at 30 fps. LabView-based home-made software was used to collect the images and analyze for displacement analysis. The optical detection system was calibrated on a certified scale with 100 μm distance marks. It was determined that 91 pixels corresponded to 100 μm length and the accuracy of the measurement technique was ±2 pixels/±2.6 μm. The software counted the difference in the number of pixels for a non-vibrating mass (initial position) and a mass vibrating at resonant frequency.

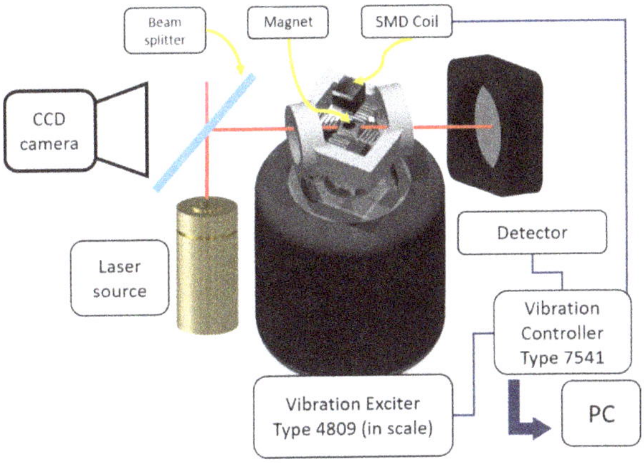

Figure 2. Scheme of the measurement set-up (optical and electrical) applied in the vibrational energy harvester's characterization.

3. Results and Discussion

All the designed harvesters were successfully printed (Figure 3) and tested. The time required for designing and printing the harvesters was less than 4 h, which is much lower than the time that would have been taken using traditional micromachining methods.

(a) (b)

Figure 3. Vibrational 3D-printed MEMS-type energy harvesters: (**a**) after printing on the printing platform before support material removal, (**b**) an example of one of the printed structures without and with an assembled miniature magnet.

The calculated Young's modulus values based on deflection vs. force curves (Figure 4), and applied in numerical simulations, were 2240 MPa for 200 μm thick and 400 μm thick microstructures. Simulations done for the data sheet modulus resulted in a lower resonant frequency, with a difference from 8 Hz to 85 Hz. The resonant values obtained for the calculated Young's modulus were closer to the real values, and the difference between the simulated and measured values was from 3 Hz to 19 Hz. The biggest difference for both the data sheet and the calculated Young's modulus were for thick (1.3 mm) springs. Thus, the Young's modulus declared by the manufacturers can be used to estimate the resonant frequency and the real (measured) value of the modulus can be used to determine the resonant frequency more precisely. Due to the various types of geometry and dimensions of the springs, it was possible to obtain resonant frequencies from 95 Hz to 340 Hz. The harvesters with circular springs had a narrower bandwidth (6–8 Hz) than the structures with "traditional" springs (20–23 Hz). This is an important observation for the future design of the printed harvesters.

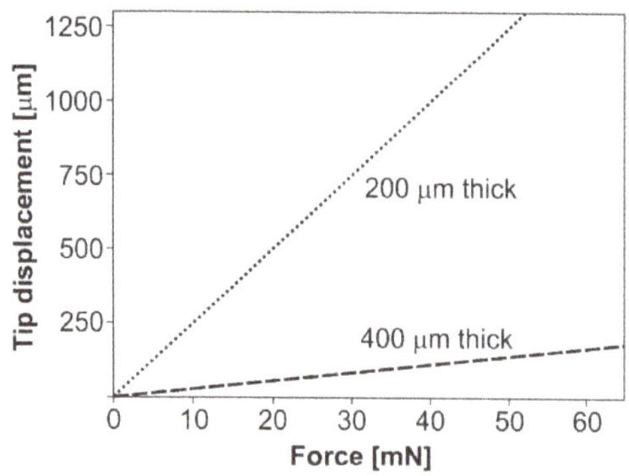

Figure 4. Examples of force vs. tip displacement of the measured 200 μm thick and 400 μm thick microbeams.

The parameters that influenced the maximal displacement were the geometry of the springs and the number of magnets. Structures with thinner springs operated properly with only one magnet. This was because of a large displacement at resonance and the sticking of the magnet to the coil. As a result, a lower output voltage and power were obtained for these structures. For a structure with two magnets and 400 μm thick springs, the maximal measured displacement of the resonating microstructure was around 1.32 mm (Figure 5).

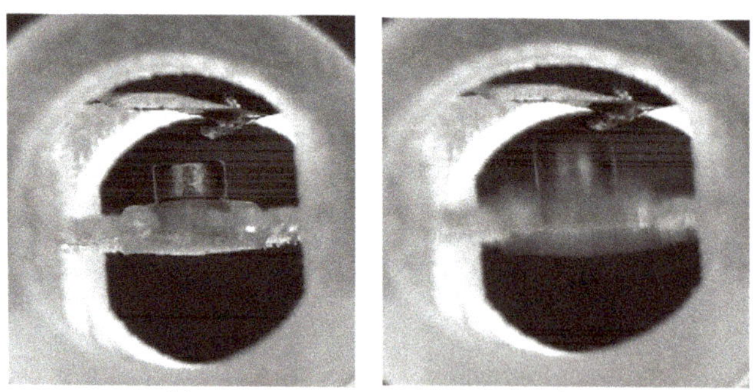

(**a**) (**b**)

Figure 5. (**a**) Images captured by the video systems at steady state and (**b**) during the resonance vibrations with 1.32 mm amplitude.

Thus, at the resonant frequency, the gap between the magnet and the coil was below 0.2 mm. This prevented these two components from sticking together during operation. On the other hand, the smallest amplitude (0.62 mm) was observed for the spiral configuration with 200 × 200 μm^2 springs and one magnet. Thus, the number of magnets (mass of the vibrating structure and the geometry of the magnetic field) and spring geometry (stiffness) significantly influenced the electrical parameters of the harvester. The maximal generated peak-to-peak voltage at resonance was 450 mV. At the optimal loading resistance (Figure 6), this was 23.7 μW output power and an energy density of around 400 $\mu W/cm^3$ for a two-magnet configuration and 400 μm thick springs. A lower value of maximal output power (12.6 μW) was obtained for a one-magnet configuration and 200 μm thick springs but, due to the lower active volume, the power density was higher (almost 600 $\mu W/cm^3$). This value is in the top range reported for other electromagnetic harvesters, and for the first time it was obtained for a 3D-printed harvester with a total volume below 1 cm^3. The calculated and measured resonant frequencies, as well as electrical output parameters, are summarized in Table 1. An important milestone in future work will be the printing of magnets together with the micromechanical structure. Some initial work in this field has been reported in the literature [38–40].

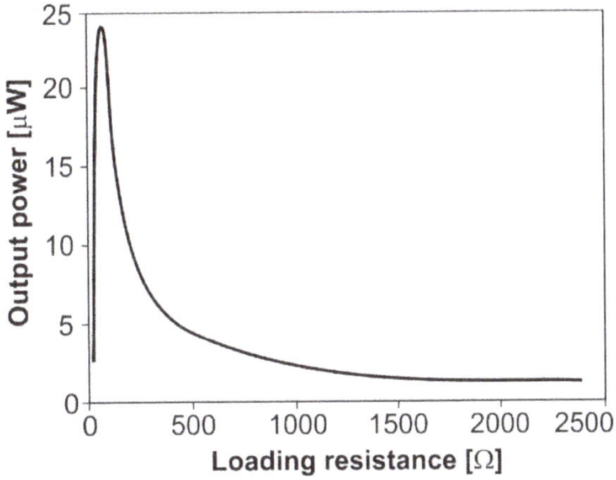

Figure 6. Output power generated by the energy harvester as a function of loading resistance, result for optimal geometry of the 3D-printed microstructure (Table 1).

Table 1. Main paramaters of the simulated and 3D-printed MEMS-type electromagnetic energy harvesters.

Visualisation of the Harvester Structure and Number of Magnets	Number of Magnets, Spring Dimensions (Height × Width [m]), and Rotation Coefficient	Simulated Resonant Frequency [Hz] for Young's Modulus 2240/1463 MPa	Measured Resonant Frequency [Hz]	Maximal Peak-to-Peak Voltage [mV]	Band Width [Hz]	Maximal Output Power [W]	Active Volume [cm^3]	Power Density [$\mu W/cm^3$]
	Two magnets, 400 × 400, 1.62	103/92	100	420	6	23.7	0.0596	397
	One magnet, 200 × 200, 1.62	98/87	95	60	8	7.2	0.0284	253
	One magnet, 200 × 200	134/117	130	254	23	12.6	0.0212	594
	One magnet, 200 × 1300	321/255	340	260	20	12.7	0.0486	261

4. Conclusions

The printed prototypes generated up to almost 24 μW of power at 1 g acceleration and 61 loading resistance. The mechanical components of the energy harvester were fabricated monolithically using inkjet 3D printing with a 200 μm or 400 μm thickness typical for MEMS devices. The total volume of the harvesters, including the frame and inductor holder, was below 1 cm^3. This was lower than other energy harvesters utilizing 3D printing to print housing or vibrating mechanical structures that have been reported in the literature. The measured resonant frequencies for various geometrical configurations were compared to simulated ones. It was found that the mechanical properties of the building material reported by the manufacturer can be used to estimate resonant frequency. More precise calculations require the real parameters of the mechanical microstructures to be determined. The 3D-printed harvesters were characterized in the measurement setup that enabled both the electrical and optical determination of the resonant frequency. It was possible to measure the output voltage as well as the amplitude of the vibrating microstructure.

Based on the results of this study, it seems that the 3D printing of MEMS-like energy harvesters is possible. Due to the short time from project to real structure, it is possible to experimentally validate real parameters of 3D-printed energy harvesters with various kinds of geometry. The performance of the harvesters we created is sufficient to power low-power sensors and electronic circuits for IoT applications. Further studies will be focused on the integration of printed magnets and multifrequency energy harvesting.

Author Contributions: Conceptualization and supervision R.W.; methodology, design, printing, and measurements B.K.; software for electrical characterization K.Ś.; validation V.C.L.; analysis and vibrational tests Q.S. All authors have read and agreed to the published version of the manuscript.

Funding: The works were realized under NAWA "Academic International Partnerships of Wroclaw University of Science and technology" program by the Polish National Agency for Academic Exchange Program (www.nava.gov.pl) and National Research Centre (project no. 2019/35/B/ST8/00687).

Conflicts of Interest: The authors declare no conflict of interest.

References

1. Dziuban, J.A. *Bonding in Microsystem Technology*; Springer Science and Business Media LLC: Berlin, Germany, 2006.
2. Esashi, M.; Takinami, M.; Wakabayashi, Y.; Minami, K. High-rate directional deep dry etching for bulk silicon micromachining. *J. Micromech. Microeng.* **1995**, *5*, 5–10. [CrossRef]
3. Shikida, M.; Sato, K.; Tokoro, K.; Uchikawa, D. Differences in anisotropic etching properties of KOH and TMAH solutions. *Sens. Actuators A Phys.* **2000**, *80*, 179–188. [CrossRef]
4. Becker, H.; Gärtner, C. Polymer microfabrication technologies for microfluidic systems. *Anal. Bioanal. Chem.* **2007**, *390*, 89–111. [CrossRef]
5. Sollier, E.; Murray, C.; Maoddi, P.; Di Carlo, D. Rapid prototyping polymers for microfluidic devices and high pressure injections. *Lab Chip* **2011**, *11*, 3752. [CrossRef]
6. Eddings, M.A.; Johnson, M.A.; Gale, B.K. Determining the optimal PDMS–PDMS bonding technique for microfluidic devices. *J. Micromech. Microeng.* **2008**, *18*, 67001. [CrossRef]
7. Unger, M.A. Monolithic microfabricated valves and pumps by multilayer soft lithography. *Science* **2000**, *288*, 113–116. [CrossRef]
8. Golonka, L. Technology and applications of Low Temperature Cofired Ceramic (LTCC) based sensors and microsystems. *Bull. Polish Acad. Sci. Tech. Sci.* **2006**, *54*, 221–231.
9. Bryzek, J. Trillion Sensors Movement in Support of Abundance and Internet of Everything. In Proceedings of the Materials of Sensors Conference, Santa Clara, CA, USA, 6 March 2014.
10. Bhattacharjee, N.; Urrios, A.; Kang, S.; Folch, A. The upcoming 3D-printing revolution in microfluidics. *Lab Chip* **2016**, *16*, 1720–1742. [CrossRef]
11. Waldbaur, A.; Rapp, H.; Länge, K.; Rapp, B.E. Let there be chip—Towards rapid prototyping of microfluidic devices: One-step manufacturing processes. *Anal. Methods* **2011**, *3*, 2681. [CrossRef]
12. Bhargava, K.C.; Thompson, B.; Malmstadt, N. Discrete elements for 3D microfluidics. *Proc. Natl. Acad. Sci. USA* **2014**, *111*, 15013–15018. [CrossRef]
13. Au, A.K.; Lee, W.; Folch, A. Mail-order microfluidics: Evaluation of stereolithography for the production of microfluidic devices. *Lab Chip* **2014**, *14*, 1294–1301. [CrossRef]

14. O'Neill, P.; Ben Azouz, A.; Vazquez, M.; Liu, J.; Marczak, S.; Slouka, Z.; Chang, H.C.; Diamond, D.; Brabazon, D. Advances in three-dimensional rapid prototyping of microfluidic devices for biological applications. *Biomicrofluidics* **2014**, *8*, 052112. [CrossRef]
15. Xu, Y.; Wu, X.; Guo, X.; Kong, B.; Zhang, M.; Qian, X.; Mi, S.; Sun, W. The boom in 3D-printed sensor technology. *Sensors* **2017**, *17*, 1166. [CrossRef]
16. Walczak, R. Inkjet 3D printing–towards new micromachining tool for MEMS fabrication. *Bull. Polish Acad. Sci. Tech. Sci.* **2018**, *66*, 179–186.
17. Rogers, C.I.; Qaderi, K.; Woolley, A.T.; Nordin, G.P. 3D printed microfluidic devices with integrated valves. *Biomicrofluidics* **2015**, *9*, 016501. [CrossRef]
18. Au, A.K.; Bhattacharjee, N.; Horowitz, L.F.; Chang, T.C.; Folch, A. 3D-printed microfluidic automation. *Lab Chip* **2015**, *15*, 1934–1941. [CrossRef]
19. Podwin, A.; Walczak, R.; Dziuban, J.A. A 3D printed membrane-based gas microflow regulator for on-chip cell culture. *Appl. Sci.* **2018**, *8*, 579. [CrossRef]
20. Song, K.; Lee, H.; Cha, Y. A v-shaped actuator utilizing electrostatic force. *Actuators* **2018**, *7*, 30. [CrossRef]
21. Hu, T.; Zhao, Y.-L.; Zhao, Y.; Ren, W. Integration design of a MEMS based fuze. *Sens. Actuators Phys.* **2017**, *268*, 193–200. [CrossRef]
22. Aspar, G.; Goubault, B.; Lebaigue, O.; Souriau, J.-C.; Simon, G.; Di Cioccio, L.; Brechet, Y. 3D printing as a new packaging approach for MEMS and electronic devices. In Proceedings of the 2017 IEEE 67th Electronic Components and Technology Conference (ECTC), Orlando, FL, USA, 30 May–2 June 2017; pp. 1071–1079.
23. Toshiyoshi, H.; Ju, S.; Honma, H.; Ji, C.-H.; Fujita, H. MEMS vibrational energy harvesters. *Sci. Technol. Adv. Mater.* **2019**, *20*, 124–143. [CrossRef]
24. Torah, R.; Glynne-Jones, P.; Tudor, M.; O'Donnell, T.; Roy, S.; Beeby, S.P. Self-powered autonomous wireless sensor node using vibration energy harvesting. *Meas. Sci. Technol.* **2008**, *19*, 125202. [CrossRef]
25. Aktakka, E.E.; Najafi, K. A micro inertial energy harvesting platform with self-supplied power management circuit for autonomous wireless sensor nodes. *IEEE J. Solid-State Circuits* **2014**, *49*, 2017–2029. [CrossRef]
26. Sano, C.; Mitsuya, H.; Ono, S.; Miwa, K.; Toshiyoshi, H.; Fujita, H. Triboelectric energy harvesting with surface-charge-fixed polymer based on ionic liquid. *Sci. Technol. Adv. Mater.* **2018**, *19*, 317–323. [CrossRef] [PubMed]
27. Sakane, Y.; Suzuki, Y.; Kasagi, N. The development of a high-performance perfluorinated polymer electret and its application to micro power generation. *J. Micromech. Microeng.* **2008**, *18*, 104011. [CrossRef]
28. Ju, S.; Ji, C.-H. Impact-based piezoelectric vibration energy harvester. *Appl. Energ.* **2018**, *214*, 139–151. [CrossRef]
29. Maharjan, P.; Cho, H.; Rasel, M.; Salauddin, M.; Park, J.Y. A fully enclosed, 3D printed, hybridized nanogenerator with flexible flux concentrator for harvesting diverse human biomechanical energy. *Nano Energ.* **2018**, *53*, 213–224. [CrossRef]
30. Seol, M.-L.; Ivaškevičiūtė-Povilauskienė, R.; Ciappesoni, M.A.; Thompson, F.V.; Moon, D.-I.; Kim, S.J.; Han, J.-W.; Meyyappan, M.; Kim, S.J. All 3D printed energy harvester for autonomous and sustainable resource utilization. *Nano Energy* **2018**, *52*, 271–278. [CrossRef]
31. Seol, M.-L.; Han, J.-W.; Moon, D.-I.; Yoon, K.J.; Hwang, C.S.; Meyyappan, M. All-printed triboelectric nanogenerator. *Nano Energy* **2018**, *44*, 82–88. [CrossRef]
32. Hadas, Z.; Zouhar, J.; Singule, V.; Ondrusek, C. Design of energy harvesting generator base on rapid prototyping parts. In Proceedings of the 2008 13th International Power Electronics and Motion Control Conference, Poznan, Poland, 1–3 September 2008; pp. 1665–1669.
33. Rubes, O.; Smilek, J.; Hadas, Z. Development of vibration energy harvester fabricated by rapid prototyping technology. In Proceedings of the 16th International Conference on Mechatronics—Mechatronika 2014, Brno, Czech Republic, 3–5 December 2014; pp. 178–182.
34. Bowers, B.; Arnold, D. Spherical magnetic generator for bio-motional energy harvesting. In Proceedings of the PowerMEMS 2008 Conference, Sendai, Japan, 9–12 November 2008; pp. 281–284.
35. Constantinou, P.; Roy, S. A 3D printed electromagnetic nonlinear vibration energy harvester. *Smart Mater. Struct.* **2016**, *25*, 95053. [CrossRef]
36. Walczak, R.; Adamski, K.; Lizanets, D. Inkjet 3D printed check microvalve. *J. Micromech. Microeng.* **2017**, *27*, 047002. [CrossRef]

37. Walczak, R.; Kawa, B.; Adamski, K. Inkjet 3D printed microfluidic device for growing seed root and stalk mechanical characterization. *Sens. Actuators Phys.* **2019**, *297*, 111557. [CrossRef]
38. Saleh, E.; Woolliams, P.; Clarke, B.; Gregory, A.; Greedy, S.; Smartt, C.; Wildman, R.D.; Ashcroft, I.; Hague, R.; Dickens, P.; et al. 3D inkjet-printed UV-curable inks for multi-functional electromagnetic applications. *Addit. Manuf.* **2017**, *13*, 143–148. [CrossRef]
39. Palmero, E.; Rial, J.; De Vicente, J.; Camarero, J.; Skarman, B.; Vidarsson, H.; Larsson, P.-O.; Bollero, A. Development of permanent magnet MnAlC/polymer composites and flexible filament for bonding and 3D-printing technologies. *Sci. Technol. Adv. Mater.* **2018**, *19*, 465–473. [CrossRef]
40. Wang, Z.; Huber, C.; Hu, J.; He, J.; Suess, D.; Wang, S.X. An electrodynamic energy harvester with a 3D printed magnet and optimized topology. *Appl. Phys. Lett.* **2019**, *114*, 013902. [CrossRef]

MDPI
St. Alban-Anlage 66
4052 Basel
Switzerland
Tel. +41 61 683 77 34
Fax +41 61 302 89 18
www.mdpi.com

Energies Editorial Office
E-mail: energies@mdpi.com
www.mdpi.com/journal/energies

www.ingramcontent.com/pod-product-compliance
Lightning Source LLC
Chambersburg PA
CBHW041955260926
51850CB00019B/2925
* 9 7 8 3 0 3 9 4 3 5 7 9 1 *